GTON

NA

AL

Cover design of 1953 & 54 Azalea Festival Souvenir Program. (Photo by Hugh Morton)

Belles & Blooms:
Cape Fear Garden Club
and the
North Carolina Azalea Festival

Susan Taylor Block

by

Susan Taylor Block

Cape Fear Garden Club
Wilmington, North Carolina

ISBN 0-9670410-3-1

Printed by Everbest Printing Co.
China

Research, photo selection, and captions by the author
Edited by Suzanne Nash Ruffin
Book and cover design by Jane Baldridge
Digital assistance by Eli Naeher

Library of Congress

First Edition

This book is dedicated to every volunteer who has ever taken part in a Cape Fear Garden Club Azalea Garden tour. Since 1953, Cape Fear Garden Club has created an annual Spring slate of beautiful spaces for the public's pleasure. Though the ticket take is lucrative, the organization channels all the profit back into the community. The club doesn't keep track of volunteer hours, nor has it sought credit for a large percentage of its gifts. Cape Fear Garden Club simply carries on the big business of flowers and the honorable practice of charitable giving, year after year.

For identifications and credits, see page 112.

Presenting the sponsors of <u>Belles and Blooms</u>:
Cape Fear Garden Club   2003

Surname	First Name	Surname	First Name	Surname	First Name	Surname	First Name
Allison	Karen	Dybvik	Christiane	Lawson	Taffy	Pyle	Linda
Alston	Linda	Eastlack	Rit G	Lee	Kitty	Quinn	Helen
Anderson	Ellen S	Echols	Hilda	Leger	Kelley B.	Rayburn	Sally
Anderson	Helen	Edwards	Ellen	Lennon	Beth	Reason	Sue W.
Andrea	Grace M	Efird	Ruth	Lensch	Terri D	Richardson	Jeannie
Andrews	Joan	Ellis	Carole W	Leonard	Carolyn S.	Rivenbark	Lola
Andrews	Teresa D	Ewing	Muriel	Lessing	Jeanie	Robinson	Lib
Arato	Addie	Faler	Bonnie L	Lewallen	Jane	Rock	Eileen
Asher	Linda	Fauser	Nancy Ellsworth	Ley	Mollie C.	Rogers	Mary Ann
Ashworth	Marie	Fedick	Mary Lou	Litchfield	Helen S.	Rose	Betty
Augustine	Carolyn	Fennell	Alma H	Little	Terry	Roseman	Patricia
Austin	Margaret	Ferguson	Charlotte Anne	Lloyd	Barbara	Rouse	Jan
Avery	Grace F	Fisher	Val	Lynch	Dianne	Saffo	Angeline
Ayers	Doris	Fisk	Joan	Lyon	Ann	Sanders	Amy
Baker	Patti	Fountain	Jean	MacLaren	Margaret (Meg)	Saunders	Bernice
Barker	Rosalind	Foyles	Annette	MacRae	Bambi	Saunders	Emma
Bauereis	Elizabeth (Betty)	Fraley	Willene	Majors	Yvonne	Schumacher	Mary A.
Beauchamp	Harriet	Friedman	Oleta	Malley	Mary	Shannon	Vera
Biddle	Madeline	Fulbright	Olive	Maready	Millie	Shaw	Babs
Bierman	Jody	Fulk	Minnie	Mathews	Betty	Shoaf	Lindy
Biggs	Susan	Gainey	Laura	Mathis	DeVonna K.	Sillars	Barbara B.
Bilzi	Sue	Gaither	Dorothy W	Mathis	JoAnne F.	Simmons	Wendy W.
Bishop	Dorothy	Galphin	Lisa P	Maus	Shannon C.	Simpson	Sandra
Bittler	Barbara	Gault	Levon	Maxwell	Cheryl Gunn	Smith	Beverly
Blacher	Martha	Geyer	Patricia	Maynard	Deborah C	Smith	Jenene
Black	Mary Ellen	Glover	Beryl S	Mayo	Gay	Smith	Karen
Block	Susan	Godwin	Cindy	McCall	Linda	Sneeden	Elnora
Bolz	Ruth-Anne	Gore	Marianne	McCall	Stephanie	Sneeden	Mary
Bowden	Elma	Gorham	Louise	McCallum	Fane	Snyder	Teresa
Boylan	Mary	Greer	Lou	McCauley	Elizabeth	Spence	Myrta
Bradley	Glenda H	Gresham	Alida	McDonald	June	Spencer	Julia R.
Bridger	Mary Ann	Griffin	Shirley	McEachern	Mary Lou	Spicer	Marilyn
Broadwater	Bettye	Grimes	Sherri	McGarry	Bitsy	Sprunt	Gloria
Brock	Jennifer	Grose	Ann	McGee	Jennifer O'Neil	Stalvey	Signa
Brown	Sylvia	Haley	Elizabeth	McIver	Gladys S.	Stellings	Pam
Brown	Thyra L	Hall	Donna	McKenzie	Kathy	Stetten	Goldie P
Brown	Tracy P	Hamilton	Bettie E	McNeir	Estelle	Stewart	Mary Lee
Brune	Ruth	Hardee	Shirley	McNeir	Hillary	Stine	Judy
Bryant	Dot	Hargy	Karin Lee	Mees	Nancy	Stokes	Lydia M
Buchanan	Barbara R	Harriss	Mary	Melton	Marty L.	Strecker	Sarah
Bullock	Jean	Hathaway	Dorothy M.	Mink	Betty	Sussman	Shaire E
Bunting	Anne	Hauge	Sally L.	Mitchell	Jane	Symmes	Nancy
Burney	Betty	Hendry	Patricia L.	Mobley	Mary	Taylor	Jo B
Burtt	Myra B	Henry	Cynthia	Monroe	Nancy	Taylor-Williams	Lilmar
Bush	Barbara	Hensley	C.C.	Moore	Brenda	Thompson	Louise
Busian	Diane P	Henson	Elaine B.	Moore	Jean	Thompson	Roxanne
Butler	Susan	Herring	Gail	Moore	Wanda	Tillet	Ruth
Byrnes	June	Hewson	Gail Tice	Mosely	MaeOmie	Toconis	Anna
Cameron	Betty	Hill	Betty	Mowbray	Charlotte W.	Tomaselli	Kate
Campbell	Mary Louise	Hill	Teresa	Murray	Leigh	Varner	Becky
Canada	Wanda	Hissam	Eleanor	Newton	Ida	Vincent	Ruth
Carter	Lorraine	Hobbs	Ginger	Nicholasen	Helen	Wallace	Betsy C
Cartier	Marjorie	Hobbs	Grace	Nix	Mary B.	Ward	Gayle L
Cassett	Marjorie	Hoover	Mae	North	Maggie P.	Warner	Alice
Chadwick	Beth	Horton	Jo	O'Brien	Sarah	Warren	Betty
Chadwick	Jo	Howle	Rita Corbett	O'Malley	Anne	Warren	Janet L
Cherry	Beth	Humphries	Andrea L	O'Neal	Glenda I.	Warwick	Jackie
Cherry	Eloise	Hunnicutt	Nell	Orski	Marcella B.	Waterbury	Jane
Christopher	Joyce	Hunt	Eleanor	Palmer	Carolyn C.	Watford	Judy
Clarke	Candus	Hunter	Melynda	Parker	Ann	Weinel	Robbie
Cline	Martha	Hussey	Alice	Parker	Connie	West	Patsy Hurst
Cook	Heather	Hutteman	Ann	Parnell	Frances	West	Virginia
Cooke	Jackie N	Hutter	Rosalie A.	Parrett	Bette	Weymouth	Peggy
Cooper	Bonnie	James	Barbara C.	Paterson	Mary F.	Whedbee	Lucile
Copley	Wanda	Jamison	Barbara	Paterson	Zafero	White	Ceclie
Coppedge	Karen C	Jarvis	Jean C.	Patz	Marilyn	Whitehurst	Mary Everett
Craft	Suzanne	Jenkins	Suzanne	Pence	Joan	Whitford	Miriam Burns
Crawley	Sherria	Johnson	Barbara	Pendergrast	Donna T.	Whitted	Lynda McIver
Curia	Rena	Kane	Barbara R.	Philbrick	Mary	Wilkins	Freda
Deaton	Rose	Kelly	Dianne R.	Phillips	Becky J.	Willard	Ann
Degnan	Gloria Jean	Kelly	Jean	Phillips	Dianne S.	Willetts	Helen
Dexter	Charlotte W	Kesler	Doris H.	Phillips	Nan	Willetts	Margaret
Dickey	Janice T	Kiner	Glenna K.	Philpott	Rebecca	Williams	Paige C
Dickhaut	Joie	King	Betty	Plaskett	Patricia A.	Williams	Shirena
Dols	Anne	King	Patricia	Porter	Wanda	Willis	Shelia
Doughty	Evelyn	Kishpaugh	Dianne M.	Pottle	Beth	Woods	Nancy J
Dreyfors	Peggy	Knox	June	Poulos	Cathy	Zigler	Nancy A
Dunajick	Millie	Lacy	Susan	Powell	Darlene	Zimmer	Roberta
Duncan	LeNeve	Lancaster	Connie J.	Price	Carol	Zimmer	Ronna
Duren-Swain	Anita	Lane	Jane	Purdum	Pauline		

Cape Fear Garden Club member Marie Ashworth supervised the creation of this magnificent float to celebrate the 50th anniversary of the Cape Fear Garden Club Tours. (Photo by Charles V. Henson)

Table of Contents

A photographic introduction:
The Bess Smith Album

The YMCA building looms on the left in this Azalea Festival parade shot taken in the 1950s by Bess Smith. Mrs. Smith, who traveled the world to photograph and tour gardens, took all but one of the photos seen here.

Bess Smith (center, front) watched the parade from the southwest corner of Third and Market streets.

Old and new: the 1839 St. James Church building and a General Electric float .

Throngs of spectators fill South Third Street and find perches around St. James Church.

The Donald MacRae House and First Presbyterian Church punctuate the skyline as the Azalea Festival parade moves down South Third Street.

A Carolina Savings and Loan Association float passes by the Colonial Apartment builidng at Third and Market streets.

Bess Smith with her sons Percy and Billy, about 1979.

Chick Mathis in the Smith Garden with a festival guest.

This photo of the Smith Garden was taken in 1953, the year Bess Smith founded the Cape Fear Garden Club Azalea Festival Tour.

ames Andrews, one of Bess Smith's gardeners, oses amid his handiwork.

Percy Smith and 1958 Orton Plantation princess Zeme North pose at a party in the Smith Garden.

Scenes in the Smith Garden.

Susan Taylor (Block), age 3, enjoys an Azalea Festival Sunday at Greenfield Lake, April 3, 1955. (Photo by George Taylor)

Introduction

Even as a very young child, I knew Wilmington was special. Unlike my cousins in the Piedmont, I was minutes from the Atlantic Ocean. The river wasn't far away either, and I could hear the sounds of ships and tugboats late into the night. And my town had a real train station instead of a wooden shed with a painted sign.

But it was the Azalea Festival with its sights, sounds, and ability to draw huge crowds of people that first showed me Wilmington was unique. In a span of just a few days, there were booming fireworks like colored drawings on a black sky; the edgy fun of the circus; and the grown-up trip to the art show at Cottage Lane. And most of all, there was the parade with all the bright floats, queens with sparkling crowns, and the bands that always included deep bass drums that you could feel in your tummy if you were standing close enough.

Today I still enjoy the festival part of our celebration, but the flower part has eclipsed the childhood joys. Azaleas, millions of them, are the real celebrities: They are beautiful in full-blown tandem and artful as individual blossoms. Even their shortcoming is an asset of sorts: The blooms last only a few weeks, but, again and again, like Christmas decorations, we look forward to seeing them — and years' worth of enjoyment only adds sentiment to a good celebration.

Every year, Cape Fear Garden Club offers up a county-wide buffet of floral treats and design delights through its Azalea Garden Tour. In fact, the club which is the oldest in the state, was 23 years old in 1948 when the Azalea Festival began and was already experimenting with garden tours. However, those early "pilgrimages" involved a collaborative effort. In 1953, Cape Fear Garden Club took on the entire responsibility of the festival tours. Fifty years later, in 2003, the Cape Fear Garden Club Azalea Festival Tour became a golden tradition worthy of a chronicle. Research has unearthed facts and testimony that prove the ties between the club and the event are even stronger than has been remembered. The young club was already on location, playing midwife, when the festival idea was birthed at Greenfield Lake. And Cape Fear Garden Club, the N. C. Azalea Festival, and Greenfield Lake intersect and reintersect through time like the tendrils of a morning glory vine climbing a white picket fence.

Belles and Blooms: Cape Fear Garden Club and the North Carolina Azalea Festival is the story of those endeavors and some of the individuals who made them possible.

— Susan Taylor Block

Ever a Gardening Spot

"On Third and Fifth Streets there are many
elegant mansions and gardens filled
with the rarest tropical and costly plants."
- Author Edward King,
writing of Wilmington, 1875

Though built on sandy soil, the Cape Fear area has sported beautiful gardens for over two centuries. By 1830, Anne Royall, author of *Mrs. Royall's Southern Tour*, wrote that the area was "the garden spot of North Carolina, and size considered, of the United States." Doubtless, the busy port, full of sailing ships from many countries, contributed to the lushness. At least as early as 1759, Royal Governor Arthur Dobbs was importing European seeds and making plans to export Venus fly trap seeds from his home, Castle Dobbs, at Brunswick Town. Plants and at least one tree that were not native to Wilmington even sprouted from ballast stones off shipping vessels that were recycled for other uses downtown.

John Burgwin, who made a good part of his hearty income from the seas, had a spectacular 10-acre garden at his country home, the Hermitage, in what is now Castle Hayne. Mr. Burgwin's friend, Captain Thomas Wright, had "wonderful gardens" at his Wrightsboro estate, Fairfield, known as a "favorite resort of the gentry."

The tradition continued in the 1800s when Wilmingtonians often bought an extra lot for the sole purpose of having a decorative garden. Platt Dickinson, Edward Kidder, James Dawson, Dr. Thomas Fanning Wood and many more all had elaborate and celebrated gardens adjacent to their homes in old Wilmington. Today, their houses are gone and their gardens are remembered only through faded photographs and the printed word.

But thanks to Emma Woodward MacMillan, author of *Wilmington's Vanished Homes and Buildings*, we have a list of plants that grew at 210 North Front Street, in the Platt Dickinson garden. A partial list includes Lady Banksia, oleander, several varieties of japonica or camellias, lilies, spirea, sweet olive, hyacinths, violets, jasmine, bay, lavender, pittisporum, lemon, orange, mimosa, mint, Japanese and wild goose plums, althea, fig, plum, sweet shrub, Siberian crab apple, old man's beard, snowdrops, and many varieties of roses such as duchess, Malmaison, and seven sisters.

In the late 1880s, one of Cape Fear's most celebrated gardens was born when Sarah Green Jones purchased an estate on Wrightsville Sound. Her husband, Pembroke Jones, named it "Airlie," and a prized portion of her original 155-acre garden remains. It is interesting that since Mr. Jones was raised by his aunt, Mrs. Platt Dickinson, he never knew life in Wilmington without a lavish garden, either on North Front Street or at Airlie. But ironically, when he later fashioned his personal

Cousins, children of Will and Engelhard Rehder, mug in the Rehder Greenhouse, 900 North 11th Street, about 1914. (Lower Cape Fear Hisorical Society)

retreat, the "Bungalow" at Pembroke Park (now Landfall), the landscaping excluded cultivated blooms of any sort. Maybe he had heard enough garden talk.

The area's most famous privately owned garden is Orton Plantation, in Brunswick County. It was still mostly rice fields in the late 19th Century, but thanks to the Sprunt family has become a nationally known and photogenic Cape Fear garden, having been the backdrop for 21 feature films and 31 television shows. Both Mrs. Henry Walters, widow of Pembroke Jones, and Mrs. J. Laurence Sprunt were members of Cape Fear Garden Club.

Enormous gardens are usually minded

MRS. N. M. MARTIN
Organizing President
1925-1927

Members of the Rehder and Schulken families parade on Nov. 2, 1891. The decorations were done by Will Rehder. (Cape Fear Museum)

by at least a few employees but, for the most part, it was left to the woman of the house to tend most of Wilmington's less famous plots. Sadly for "yard men," labor was sinfully cheap. Residents with an average income could easily afford some help in the early 1920s, but many a woman enjoyed the work. As an almost custodial duty she planted and watered, peered and prayed, and finally exulted at the brightly colored blooms that an unseen force opened beneath her bedroom window overnight.

The Club

At its core, the Cape Fear Garden Club actually began as a small literary roundtable known as the Tuesday Book Club, organized in 1923. It met at the Y.W.C.A., located at 206 North Second Street, about where the entrance to the library parking deck is now. But on January 20, 1925, members Mrs. Sudie Casper, Mrs. N. M. Martin, and Mrs. Martin Willard said that they were more interested in discussing gardening than books. They proposed the club's purpose be changed.

The organizational meeting of the Cape Fear Garden Club was held February 11, 1925, in the Great Hall of St. James Church.

St. James Church, about 1919 (Cape Fear Museum)

There were twelve charter members: Mrs. Nathaniel Macon Martin (president), Mrs. John Bolles, Mrs. Platt Davis, Mrs. W. E. Elliott, Mrs. W. A. Graham, Mrs. A. M. Hall, Mrs. R. H. Hubbard, Mrs. William Latimer, Mrs. Sarah Lippitt, Mrs. Hugh MacRae, Mrs. Jeanie Strange, and Mrs. Martin Willard. Each was encouraged to recruit five more members.

The Dram Tree. (Lower Cape Fear Historical Society

Determined to have a distinctive gavel, the ladies somehow procured a branch of the Dram Tree, the ancient landmark that used to be the watery fork in the road to sailors going either to Brunswick Town or to Wilmington. A gavel was carved from the branch and a tiny plaque noted its origin. The Dram Tree existed as least as late as 1939 and drew its name from the tradition of sailors taking a dram of rum either to celebrate their passage through the tricky shoals of Cape Fear or to steady their nerves as they headed for the open seas.

Quaint etiquette led the women to use their husbands' names, but in many ways they were ahead of their time. Determined to preserve the environment and bring about increased awareness of nature's beauty, they wrote a new Club Objective: "To promote the cultivation of flowers, shrubs, and trees and thereby to beautify the homes, streets, and highways of the community, that this beauty may enrich the lives of all." Eventually the stated purpose underwent some editing, to include the preservation of native birds.

Carolina yellow jessamine, a long-time

MEMBERS OF THE CAPE FEAR GARDEN CLUB

————

MRS. PERCY ALBRIGHT
114 South Third Street   Telephone 357
MISS LILLA BELLAMY
515 Market Street   Telephone 245
MRS. JOHN D. BELLAMY, JR.
323 South Third Street   Telephone 577-J
MRS. E. M. BEERY
408 Market Street   Telephone 180
MISS LENA BEERY
125 South Fifth Street   Telephone 2113
MRS. E. K. BRYAN
11 South Fifth Street   Telephone 2013-W

MRS. JOHN CAMPBELL
316 South Third Street   Telephone 736
MRS. R. C. CANTWELL, JR.
Oleander   Telephone 1413-R
MISS MARY LUCAS CANTWELL
8 South Sixth Street   Telephone 458-W
MISS ELIZABETH CHANT
Cottage Lane
MRS. O. F. COOPER
106 North Sixth Street   Telephone 1714-J
MRS. J. B. CRANMER
311 Market Street   Telephone 1424
MRS. FRANK CRANE
Oleander   Telephone 2080-J
MRS. JOHN D. CORBETT
1710 Market Street   Telephone 918

MRS. PLATT DAVIS
224 South Third Street   Telephone 1998-W
MRS. MARSDEN DeROSSETT
Oleander   Telephone 1982-J
MISS JANIE DUNN
104 North Seventh Street   Telephone 1613-J
MRS. ARTHUR DUNNING
Greenville Loop   Telephone 7523-W

MRS. W. G. ELLIOTT
105 South Fifth Street   Telephone 1997-J

MRS. L. E. FARTHING
20 South Fifth Street   Telephone 377

This page from the 1928 Cape Fear Garden Club Booklet includes the names of many "old Wilmingtonians," as well as noted artist Elizabeth Chant. (New Hanover County Public Library)

Southern favorite, was chosen as the club flower. A harbinger of spring, the flower resembles a butter-colored honeysuckle. Though poisonous in plant form, it is now a vital part of some nerve medications.

By 1928, the group had grown to 80 members and had its first yearbook, a tradition that continues today. The club also named honorary members: Mrs. Henry Walters of Airlie and

The 1931 garden club yearbook included some charming sponsorships like Belk-Williams' advertisement. Others included sunburn lotions, marketed by Adolph G. Ahrens, and a Wilmington Stamp and Printing Company advertisement that read, "Beautiful gardens and good printing are synonymous." (New Hanover County Public Library)

New York; Mrs. S. C. Damon of Kingston, Rhode Island; Mrs. Nathaniel Macon Martin, honorary president; and Mr. T. P. Lovering.

In 1929, members of Cape Fear Garden Club agreed to align with The Garden Club of N. C., Inc., a younger organization. Two local club presidents would later become state presidents: Mrs. J. B. Cranmer and Mrs. J. Buren Sidbury. Federation with the national garden groups occurred in the early 1930s, but Cape Fear's ties within the state have always been more cherished.

Though the club brought in speakers from other places, most often N. C. State and the University of North Carolina, they also could find enough local authorities to fill up a yearly calendar. In 1932 and '33, the ladies

engaged a slate of speakers with familiar names: Ludeke, Boet, Rehder, Tinga, Verzaal, Roudabush, Lucy B. Moore, and representatives of T. W. Wood and Sons.

In the early 1930s, Mrs. E. K. Bryant opened her garden at 11 South Fifth Avenue for the first known Cape Fear Garden Club tour. Only members were invited, but the concept took root in some of their minds. In 1942, Claude Howell, whose mother was a member of the club, drew his interpretation of the beauty of Mrs. Bryant's garden as a club yearbook cover.

Cape Fear Garden Club yearbook, 1942-43. Cover by Claude Howell. (New Hanover County Public Library)

(New Hanover County Public Library)

A Trailblazing Member

In September 1933, a particularly vile hurricane swept by Wilmington on its way to Carteret County. The residue it left on Wrightsville Beach awakened interests in one garden club member, Mrs. Cecil Appleberry, that would have far-flung results. Mrs. Appleberry, whose first name was Edna — but seems always to have been known by everyone by her melodic married name — walked the beach after the storm. According to local author and history compiler Bobbie Marcroft, who interviewed her in 1977 for *Scene* magazine, Mrs. Appleberry said she found "the front of the beach piled about two feet high with animals – some of them nobody on the beach had ever seen and I was absolutely fascinated

and frantic because I didn't know what they were."

Seagate School, crafted by the same talented neighborhood residents who built at Wrightsville Beach, Airlie, and Pembroke Park, sports an Oceanic-style dome. (New Hanover County Public Library)

She set up a shell display case at Seagate School and began making "as many as five talks a day" in area schools. As a result, Mrs. Appleberry's fame grew, by conversation and in print, statewide. Cape Fear Garden Club gave her plenty of support and member Rena MacRae's husband Hugh asked Mrs. Appleberry to create a live sea animal exhibit at Lumina, the oceanfront pavilion he owned through his company, Tidewater Power. The exhibit lasted ten years.

During that period, according to Mrs. Marcroft, the naturalist took over the Wrightsville Beach ABC store and turned it into an off-season exhibit, "Appleberry's Marine Museum." The ABC store, located on the northeast corner of North Lumina Avenue and Stone Street, used to be closed about nine months of the year, and the shelves created perfect spaces for exhibits. Mrs. Appleberry stocked the original shell show as a Cape Fear Garden Club project— a sort of marine hospitality house for the N. C. Garden Club that was meeting in Wilmington. Locals could also browse but had to pay a nominal admission fee that Cape Fear Garden Club contributed to St. Andrew's on-the-Sound.

Despite the fact that the liquor had been stored elsewhere, the building still said "ABC." It brought chuckles from visitors, and was nectar for at least one beach busybee. Two women were taking a stroll down the old boardwalk and passed the museum door just as Mrs. Appleberry was taking a sip of soda. "Just look at that brazen hussy drinking in broad daylight," one woman told the other. The shell collector delighted in telling the story for years to come.

Later, in 1942, aging Wilmingtonian Theodore Empie, donor of Empie Park, charged Mrs. Appleberry to take his place as Wilmington's leading bird enthusiast. Though she accepted with some reluctance, she soon founded the Wilmington Bird Club, forerunner to the Wilmington Natural Science Club. By 1950, when she chaired Cape Fear Garden Club's Conservation Group, visitors from many states came to meet her. They wanted to hear her lectures and to go on bird sighting trips in her company. Her "sermons" on the importance of caring for bird populations and her bird counts were known by ornithologists all over the country.

In 1950, she recorded 184 species of birds. At Greenfield Lake she sighted, among other species: parula, prothonotary, yellow-throated warbler, Arcadia flycatcher, orchard oriole, cardinal, American and snowy egret, fox sparrow, great horned owl, osprey, bald eagle, great blue and green herons, European widgeon, Florida gallinule. At Airlie, she noted that painted buntings were abundant. At Orton, she found warbler, painted bunting,

Mrs. Cecil Appleberry (left) addresses Mrs. P. R. Smith (right) at a Cape Fear Garden Club Conservation Committee meeting, in 1951. According to Hugh Morton, Mrs. Appleberry lobbied for state laws that restricted commercial shipment of Venus flytraps. This photo, by Mr. Morton, first appeared in *The State*, March 1, 1952. (New Hanover County Public Library)

woodpecker, pileated woodpecker, osprey, water turkey, and gallinule.

In 1952, with the help of Mrs. C. D. Maffitt, she opened a nature museum in the old greenhouse at Greenfield Lake. Noted as a feature of the Azalea Festival that year, the museum highlighted not only birds but sea and plant life. It was manned by the Wilmington Natural Science Club, many of whose members were also members of Cape Fear Garden Club. The clubs shared speakers, including the world-renowned ornithologist Alexander Sprunt, Jr., of Charleston.

Mr. Sprunt's cousin Kenneth Sprunt remembers Mrs. Appleberry fondly. "In my early youth," said Mr. Sprunt, in 2001, "I was given a paddle boat to use in Banks Channel at Wrightsville Beach by my father. I think he paid five dollars for it. I paddled this little boat all around the sound. There wasn't much traffic in the late twenties. I found all these creatures that grew in the water. I didn't know exactly what they were or what they stood for.

"It just so happened that Mrs. Cecil

Appleberry lived behind Lumina in one of several apartments called Pomander Walk. Before Waynick Boulevard was there, there was a series of small cottages there and the Appleberrys spent the summer there. Mr. Appleberry worked for the Coast Line. I knew Mrs. Appleberry and I knew she knew all about birds and animals and creatures, and I would go to see her and take these creatures I had gotten from the waters on the sound. She was a good many years older than I was, but she was a wonderful lady. That's why I developed my interest and love of all things that wiggled and swam and flew and so forth.

"Many years later, she was in charge of and founded the Wilmington Bird Club, the Wilmington Natural Science Club and as such she was a frequent visitor down at Orton where

Orton Plantation. (Photo by Hugh Morton)

we have a lot of birds and things, and one day she called my wife, Betsy, and said, 'Betsy, have the martins arrived yet?'

"Well Betsy, who was in the midst of polishing floors with Louis Orris, was thinking, 'Dear Lord, there's a whole lot of people named Martin.'

"She said, 'No, I haven't seen them.'

"Mrs. Appleberry said, 'Well they were here a little while ago and I expect they'll be

down there pretty soon.'

"So Betsy quit her work, changed clothes, and made coffee. Sure enough, Mrs. Appleberry showed up in about an hour. Betsy was looking all around and didn't see anybody named Martin or anybody else. But Mrs. Appleberry was looking up in the trees— looking for a different kind of martin."

"'Of course,' said Betsy. 'Martins are the first birds of spring!.'"

Big Doin's

On May 12, 1933, Cape Fear Garden Club held its first Flower Show in the Sunken Gardens at the Courthouse. The second Flower Show, held a year later was a bigger success. Mayor Walter Blair opened the show, held at St. James Church in the spring of 1934. Every room of the church's parish house was used for displays created by various members. Gardeners sold potted plants and refreshments to benefit various projects. Beloved florist Will Rehder had a lavish display, and his only real competitor, Lucy B. Moore, filled a stage with her wares: a sundial, arbor, and a garden party mannequin. Dr. Charles Dearing was one of the judges. The show was overwhelming, both in beauty and variety. Mrs. Arthur C. Diehl, chairman, said, "It was our first – nobody knew what to leave out!"

In 1936, Cape Fear Garden Club spawned another group: the Crepe Myrtle Garden Club. In their first year, the new club planted over 200 crepe myrtle and dogwood trees throughout Winter Park. The Crepe Myrtle Garden Club would be the first of many clubs generated from Cape Fear.

In 1939, Cape Fear Garden Club hosted the first Camellia Festival under the leadership of Allie Morris Fechtig. Held in the Great Hall of St. James Church, it was an elaborate show that spawned annual encores. The 1941 camellia show featured "Living Pictures,...a series of beautiful tableaux created by Mr. Henry MacMillan."

Hazel Knight's wedding decorations. Will Rehder was a local pioneer in special occasion design. (Cape Fear Museum)

Henry Burbank Rehder was a leader in the forerunner to the Azalea Garden Tour: the Azalea Pilgrimage Group. A grandson of Johanna Rehder, founder of the oldest florist enterprise in North Carolina, he was born into a rich tradition. His family helped decorate for the famous wedding of Mary Lily Kenan and Henry Flagler in Kenansville, the 1912 wedding of Sadie Jones to John Russell Pope at Airlie, the reception given by James Sprunt

for President Taft, and many other notable events.

The desire to share the beauty of flowers was second nature to Henry Rehder. According to him, in 1944, Cape Fear Garden Club helped to initiate the forerunner to the Azalea Garden Tour.

"Along with the N. C. Garden Club," said Mr. Rehder, "we recognized the fact that Wilmington had so many beautiful gardens that people wanted to see — but how could they get access? So with the assistance of Cape Fear Garden Club, we formed the Azalea Pilgrimage Group. But it wasn't exclusively a garden club project. It was made up of about 40 gardens and businesses and sometimes it was hard to line up that many places."

According to Mr. Rehder, some of the pilgrimage stalwarts were Allie Morris Fechtig, Daisy Page, Marie Gerdes, J. B. Cranmer, Bess Smith, and Annie Gray Sprunt. Henry Rehder and Marie Gerdes, president of Cape Fear Garden Club in 1944-1945, were cousins and worked closely to make the Pilgrimage a success. Mrs. Sprunt hosted a luncheon at Orton for Pilgrimage and for many years for the garden club tour. Cape Fear Garden Club helped with the lunches and young girls were recruited to be greeters. Airlie was usually included on the tours.

"C. Mac. Davis," Mr. Rehder continued, "who was president of the Atlantic Coast Line Railroad, was a supporter. He had asked local nurseries to contribute to the Atlantic Coast Line Garden between the freight traffic department building (now the Wilmington Police Department) and Bridgers Trading Co. Will Rehder Florist created the design and maintained it for many years. It was full of formosas, and had a trellis and garden seats. ACL employees often had their lunches there. The Coast Line Garden was on the Pilgrimage Tour."

Bess Smith and Marie Gerdes, both so active in the Pilgrimage group, also master-minded a new Cape Fear Garden Club project in 1945: Holiday House. Held at St. James Church, the event attracted great numbers of people who wanted to see artistic ways to "deck the halls" of their homes, trim a tree, and wrap their Christmas packages. Emma Gade Hutaff and Mrs. J. M. Gregg provided music and garden club members served refreshments. But the event lasted only one afternoon, and positive word-of-mouth left many disappointed that they did not attend.

For the next several years, Holiday House was held at the American Legion building, originally the Bridgers home and now Graystone Inn. Thousands of guests enjoyed an elaborate show of Christmas decorations that filled the entire first floor and the stairway of the building.

Cape Fear Garden Club, along with *Star News* publisher Rye Page, and the Wood and Cross seed stores, helped support the Carl Rehder School Garden Contest during the late 1940s. Prize money was awarded to about 350 students a year for producing impressive vegetables, fruit, or flowers from home gardens.

In 1948, the objectives of Cape Fear Garden Club were beach, roadside, plaza, and park beautification, as well as influencing city fathers to create truck lanes, for omnipresent soot blackened the homes on South Third Street and soiled the very air. Harder yet, members of the club waged a battle to save trees along Third Street, oaks threatened by the city's program to widen roadways used by the large and noisy trucks. Sue Hall, a Wellesley graduate and a member of Cape Fear Garden Club, even wrote a poem in 1950 titled "Traffic: A Protest over Third Street." It included the following lines:

"Don't you care for ancient beauty
Long deep-rooted in our soil –
Trees that generations sheltered,
Sons of rest and sons of toil?

"Don't you like to think and ponder
On the values quite unseen
That have made our City different
In the centuries between…

"We cannot hear the church bells chime,
Nor lilting song of birds;
Conversations are frantic shouts,
That once were pleasant words….

"Get a move on! Heave the axe!
Just a jiffy it will take
To chop down what our Father God
Took two hundred years to make."

Into the 1940s and '50s, the literary nature of the parent club was still evident. Words by Kipling, Cowper, and Thomas Hardy made their way into the yearbooks and occasionally original verse was used to promote garden club events.

The haunting beauty of Greenfield Lake, about 1929. (Photo by Henry L. Sternberger, Cape Fear Museum)

Greenfield Lake

The garden club continued to consider Wilmington its personal landscaping project, planting live oaks at James Walker Memorial Hospital, introducing shrubs and flowering plants to Post Office Park at Second and Chestnut streets, and beautifying the grounds of many area schools. But if Cape Fear Garden Club had a heartstring in those years, it was Greenfield Lake. A natural wonder, the lake was once part of a plantation owned by Dr. Samuel Green. Known as Green Fields, the mill pond was considered far from the city. The property went through several owners before it became known as McIlhenny's Pond,

named for owner Thomas Cowan McIlhenny, who rented rowboats there from 1900 until 1924. According to Cape Fear Garden Club member Helen Weathers McCarl, "Many a young man courted his sweetheart among the water lilies and cypress trees."

Then Greenfield lovers and lovers of Greenfield were horrified when a carnival business purchased the lake, in 1924, and walled it off with a high board fence. Campers were encouraged to rent tiny huts and outhouses dotted the land, but few responded. The entire area soon looked like a littered and abandoned campsite. In October, a fair was held there and members of the Art League of Wilmington hung an exhibit in one of the huts. Mrs. Rufus W. Hicks, North Carolina Sorosis leader, disgusted with the carnival owner's negligence, painted a sign and placed it on the outside wall of the cabin: "Why not make a park of this beauty spot?"

The ladies' industrious labors were not lost on the owner. Soon he began clearing away debris, razing bad construction, and dismantling the wooden fence. Then he turned the entire area over to the N. C. Sorosis, a jubilant crowd that immediately began working overtime to transform Greenfield. They painted benches, planted flowers, created a tearoom in the old pavilion, and organized barbecues and concerts to raise money for more improvements.

When the next city election rolled around in 1925, voters approved the purchase of Greenfield Lake, and the carnival owner walked away with $65,000. Commissioner of Public Works J. E. L. Wade, gardener Carl Rehder, and Louis T. Moore, chairman of the Wilmington Chamber of Commerce, led a new campaign, encouraging businesses and individuals to donate plantings to the garden. As a result, the area experienced rapid enhancement. In addition to the many individuals who donated one or two shrubs, T. J. Armstrong of Pender County gave an entire palmetto thicket and George Trask delivered truckloads of flora.

The city provided lights, paths, colored illumination for the old millpond spillway, goldfish ponds, rock gardens, rose arbors, stone benches, new bath houses, playground equipment, tennis courts, a zoo, and fish. The infant organization, Cape Fear Garden Club, contributed plants and cash, and made strong emotional ties to the space many continued calling the city's "beauty spot." However, visitors could enjoy it only from South Third Street. There was no road around the rest of the lake.

Nevertheless, things were going well until 1929, when the Great Depression made bread more important than blossoms. By 1931, many husbands and fathers who once had good desk jobs, had lost their positions, homes, and happiness. Some of Wilmington's more economically comfortable citizens couldn't even scrounge up enough money to buy shoes. Several local businessmen committed suicide.

A citywide program used the donation of one-day-a-month's pay from contributing workers to employ 1500 men to build a five-mile road around Greenfield Lake. Even with the modest wages of the Depression, Wilmingtonians raised an admirable amount:

Greenfield Lake at the peak of its beauty. (Photo by Hugh Morton. Cape Fear Museum)

$110,000 over a period of eighteen months. Additionally, the government sent prisoners to help. To honor the generosity of the steadily employed, the road was named Community Drive.

Even in the winter of the Depression, flowers gave hope and a symbol of so much more to many Wilmingtonians. Poet, professor,

19

Many of the azaleas that border the lake were transplanted from club members' gardens. (Photo by Hugh Morton)

The Seeds are Sown

At last, the millpond had metamorphosed. Greenfield's colorful new existence was recognized statewide in 1932 when the N. C. Federation of Women's Clubs voted it the "Beauty Spot of the State." And from lakeside allure sprung the idea of the Azalea Festival. One Sunday afternoon in the spring of 1934, Dr. Houston Moore, a Wilmington physician, and his wife were motoring down Community Drive, admiring the peak of azalea blooms. The couple saw the potential for enhancing the lake's beauty and Dr. Moore suggested the idea of a festival that would coincide with the narrow window of the azaleas' flowering season. Despite the bleakness of the Depression, he could almost see floats and bands, flower shows and beauty queens.

and novelist Jessie Rehder, Henry's sister, wrote the following poem to her cousin, Marie Rehder Gerdes — one garden lover to another.

"A solitary starling, black in flight,
Goes down the sky; the air is like a sword
Against my frozen cheek, and in the night
Not one faint echo of a friendly word.
The apple trees, in winter bare design,
Their leafless branches stark as prison bars,
Refute the promise of a day divine,
Shout down the song implicit in the stars.
But deep within the stiff, macabre boughs
The sap already risen, and with spring,
Leaves will be green and gay as lovers' vows,
Fruit will form again, and birds will sing.
Thinking upon the dress that April wears,
I can forget the winter and its fears."

—Jessie Rehder, Christmas 1937

As a direct result of the Depression, a landscape architect was hired to design new improvements. Funds came from federal and local government to employ Charles Freeman Gillette (1886-1969) of Richmond. An apprentice under Boston landscape architect Warren Manning, Gillette had a distinguished career that spanned more than 50 years and included notable work at the Virginia executive mansion; Woodrow Wilson's birthplace in Staunton, Virginia; Washington and Lee University; and the University of Richmond. According to Hugh MacRae II, the landscape architect also designed the garden at the Bugg-MacRae House at 807 Forest Hills Drive and at the home of Col. George Gillette (MacRae's stepfather), a relative of the landscape architect, who lived at 1740 Live Oak Parkway. Charles Freeman Gillette also did some street designs for the Oleander Company, Mr.

MacRae's development firm.

Charles Gillette supervised the creation of cedar-lined entrances, cypress guard rails, and rustic footbridges. Mr. Gillette also sloped the lake's banks and planted ivy and honeysuckle on them. He added picnic tables and benches where they would afford the best views. And he planted quince, camellias, spirea, gardenias, dogwood, yaupon, rambler roses, yucca, magnolias, crepe myrtles, Venus fly traps, swamp plants, ferns – and more azaleas.

In 1936, Dr. Moore shared his beautification and festival ideas with a group of local leaders that included several garden club members who met at his office at Greenfield Lake. The aesthetic improvements pursued were to some part realized because, again, many Wilmingtonians dug up azaleas from their own yards and replanted them at Greenfield Lake. The festival was tabled, though, because of continued concerns about the economy. However, Dr. Moore had succeeded in creating a secure network for the future, comprising Cape Fear Garden Club, Greenfield Lake, and the Azalea Festival.

Dr. Moore's dream of Greenfield beautification did go forward, if slowly. An earlier campaign had raised $2500 for improvements. When the money ran out, Dr. Moore himself briefly hired a man as a fulltime gardener. A second campaign, spearheaded by Dr. Moore, brought $2,000, and the city agreed to provide two workers and several prison units to make the money go further. Another lake enthusiast, John Spillman, actually worked alongside the paid laborers, volunteering his gardening services

for two years.

In 1938, the Greenfield Drive Association was formed. Dr. Houston Moore was chairman and Allie Morris Fechtig was Cape Fear Garden Club's representative. Miss Fechtig gave generously to help improve the new road and add landscaping. And most garden club members gave azaleas from their own yards to replant around the lake.

The following year, on August 30, 1939, the Greenfield Association was incorporated by Mrs. Carl Powers, Miss Allie M. Fechtig, Louie E. Woodbury, Jr., Bernard S. Solomon, Earl Napier, John Spillman, Jr., Mrs. B. M. Jones, H. Churchill Bragaw, Emsley A. Laney, Mrs. J. C. Williams, Alan A. Marshall, Mrs. C. D. Maffitt, H. R. Gardner, and J. E. L. Wade. The group agreed to purchase plants from Orton Plantation, but the Sprunt family donated many more than were bought. On the advice of Will Rehder, various groups like

Historian Elizabeth F. McKoy created this 1771 model of Wilmington based on a 1769 Claude Sautier map housed at the British Museum. Miss McKoy crafted the model in 1939 as part of the 100th anniversary celebration of the 1839 St. James Church building. According to Miss McKoy's niece, Elizabeth McKoy McCauley, she joined Cape Fear Garden Club, entered the model in a statewide garden club project competition, and won the event. The model was exhibited at City Hall and Cape Fear Museum for years. It was only natural that Miss McKoy would be interested in buildings: Lincoln Memorial architect Henry Bacon was her uncle. (photograph by Elizabeth McKoy, courtesy of Elizabeth McKoy McCauley.)

Rotary, Kiwanis, and the Lions Club adopted sections of the lake's shore. Allie Fechtig led Cape Fear Garden Club in fixing up a parcel of land. Through successive years, the club took on several areas.

World War II altered Cape Fear Garden Club's activities and delayed plans for further improvements to Greenfield and the formal organization of a flower festival. Cape Fear Garden Club held fewer meetings and the programs were generally more somber. In 1943, Archibald Rutledge, South Carolina's Poet Laureate, spoke to the ladies on the horrors of war and its effect on flowers and wildlife. Officers' wives were invited to all club meetings and members sometimes gave them guided tours of some of the area's best gardens. Most contributions went to the American Red Cross, located at 411 South Front Street. In addition, garden club members eventually planted the Red Cross headquarters lawn with flowers and shrubs.

Parting the Gates

By 1945, Cape Fear Garden Club was taking part in a statewide North Carolina Garden Club promotion: Spring Pilgrimage Tours. But Wilmington became the proverbial "tail wagging the dog" because of its strong club, organizers like Henry Rehder, and its vast natural garden resources.

After the war, Wilmington's economy sagged. Emptied of about 130,000 servicemen and shipyard workers who had left by late 1945, the city needed a financial boost. New tourist promotions were created to tout

Garden Club president Marie Rehder Gerdes (on left) presides over 1945 meeting in which it appears blessed good news is being delivered. (New Hanover County Public Library)

Greenfield Lake and, during 1946 and 1947, workers added 175,000 plants to the area. It was in 1947 that Hugh Morton and Dr. Houston Moore began in earnest to plan a festival. A group of interested persons, representatives of various civic clubs, began to meet. The first group included R. W. Snell, Hugh Morton, T. T. Hamilton, C. C. Johnson, L. C. LeGwin, Joe Ray, Stanley Rehder, Guerard Simpkins, Mrs. Walter E. Curtis, Verna Sheppard, Dr. W. H. Moore, Kenneth M. Sprunt, Mrs. Lewis L. Merritt, Mrs. Ransey Weathersbee, Mrs. W. A. Fonvielle, and Mrs. Clayton

Margaret Tannahill Hall, first director of St. John's Art Gallery, created a depiction of Airlie Road for the 1946-47 Yearbook Cover (The Louise Wells Cameron Art Museum)

Lots of woman-hours went into this dramatic transformation. Cape Fear Garden Club decorated the old U.S.O. building for its Camellia Show, January 17, 1947. By that time, the structure, located at 126 South Second Street, was known as the Community Center. (New Hanover County Public Library)

Grant.

At one of the early meetings, members elected Hugh Morton president in his absence. At the next meeting, Mr. Morton tendered his resignation but Dr. Moore refused to accept it. Morton then began to draw plans for a small event that would include a "parade, dance, 'community sing,' the bringing of a Hollywood star to serve as queen, and a flower show." A representative of Cape Fear Garden Club attended the festival's organizational meetings

Cora Preston, Bess Smith, and Marie Gerdes (left to right) had enough decorations at home to dress up the Community Center for the 1947 Spring Garden Show. (New Hanover County Public Library)

and made recommendations to both to the club and the Cape Fear Garden Council concerning opportunities for service and monetary contributions.

1948

Finally the Azalea Festival became official when local attorney Wallace Murchison drew up incorporation papers for the event. Cape Fear Garden Club was one of 36 "Incorporators of the Azalea Festival." According to the March 28, 1948, edition of the *Sunday Star News*, others included the Wilmington Chamber of Commerce, various civic clubs, Airlie, Orton Plantation, Greenfield Drive Association, several fraternal lodges, and the "Homeowners' Association of South Third Street."

Dr. Houston Moore, who would live just long enough to see his dream come true, must have been thrilled with the way Hugh Morton conducted the first tour. Morton, a representative of the Jaycees in the incorporation process, handled the event like a seasoned advertising executive.

"I told people we were the city of over a million azaleas, and that was just among Orton, Airlie, and Greenfield Lake," said Mr. Morton, who also encouraged individual homeowners to plant azaleas. Garden expert Jim Ferger concurred, "In no other section of the South should the display of color made by Azaleas surpass what we can attain in our Wilmington area." The result was a natural fireworks display that lit up the entire city.

"Unlike Charleston, where they had their own azalea festival, we planned the festival to coincide with the peak of the blooming season," said Hugh Morton, in 2003.

"Charleston's festival died away, but the Azalea Festival got stronger." It also didn't hurt Wilmington's chances that Charleston azaleas experienced a blight before 1948.

"We started small that first year," continued Mr. Morton, "because I knew if we didn't show a profit, the whole idea would fail." When treasurer Kenneth Sprunt totalled everything up, the officials were pleased: The first festival turned a $5,000 profit, a lot of money in those days.

Jo Beatty Chadwick, who grew up near Beatty's Bridge in Bladen County, remembers the small-scale charm of that first festival. Her wedding was planned for exactly two weeks after the festival and, in anticipation, some friends gave her a luncheon at St. John's Tavern, 110 Orange Street, the day the festival began. "We knew there was something going on, but it didn't interfere with our plans. It was the morning of the parade and my father drove us downtown, all the way to St. James Church, to let us off. My mother and I stopped and stood on the steps of St. James Church to watch the parade for a few minutes before we walked to St. John's Tavern. There we were in dining room and the Queen and Hugh Morton and the other officials were having lunch in the other room. It was a small festival."

The first parade was, indeed, a modest one. Though variously reported as anywhere between eleven and 31, it actually contained 16 floats. Henry Rehder, float chairman, said they originally aimed at having just one, but there was enormous enthusiasm. Mr. Rehder "appealed for large flat-bed trucks around which to build parade floats," said the *Star News*, March 29, 1948. With rough-textured tractors pulling the floats, the big-city beauty queens framed in fluff and flowers appeared

A tractor pulls a float by Union Station, at Front and Red Cross streets. (Cape Fear Museum)

even prettier by contrast. The floats included one bedecked with pirate girls and one sponsored by the florists of Wilmington that featured Beth Harriss, daughter of a festival organizer, David Harriss.

The Garden Party Begins

It was actually the garden clubs' Flower Show that launched that very first Azalea Festival, at 11:30 a.m., Friday, April 9, 1948. Broadcaster Ted Malone covered the opening live for ABC Radio News to an estimated 10 million listeners. Mr. Malone, known coast to coast for his commentaries and news stories, brought a lot of attention to the festival and attended for several years. William Burns and his brother, Jim Burns, had both worked at ABC in New York with Malone. It was Billy Burns who persuaded him to visit Wilmington during the festival. While here, Malone had a fine local historian as an escort: author, lecturer, and photographer Louis T. Moore.

Mr. and Mrs. Ted Malone (on left), Billy Burns, and Azalea Queen Jacqueline White posed after Mr. Malone's first nationwide broadcast of the festival's opening event: the Cape Fear Garden Club Flower Show. (Photo by Henry L. Sternberger, Cape Fear Museum)

Center, on the northwest corner of Second and Orange streets. The opening event was a joint effort. Cape Fear Garden Club joined forces with the N. C. Sorosis Garden Club and the Crepe Myrtle Garden Club to create a show that delighted visitors until 10 p.m., and continued throughout the day, on Saturday. The show was not merely a display of perfect blooms, it was an elaborate exhibit featuring a mock outdoor-style formal garden, a beautifully appointed bride's table, and a miniature "Greenfield Lake" bordered by ballast stones.

In addition to hosting the annual garden pilgrimage, members of Cape Fear Garden Club also helped publicize the festivals' "Recommended Garden Tours," composed of Airlie, Greenfield Lake, and Orton. Working with Gregg Bros., a local company, they produced an oblong wooden Burwood bowl to promote the tours. Carved images of Greenfield Lake and Airlie and Orton mansions highlight the bowl, and azalea blossoms decorate the borders.

Arrangements even were made for one physically handicapped garden lover to have special tours of Airlie and Greenfield in an ambulance. A song festival at Legion Stadium featured bands and singers from local schools, with the performers from Williston bringing loud applause. There was also a military band concert at Greenfield Lake.

The garden club named Mrs. Rinaldo B. Page honorary chairman. Daisy Page was an avid gardener who transformed her home, Turtle Hall, from a plain and simple soundfront estate to a lavish garden of azaleas, camellias, and exotic trees she brought home from her trips abroad. As the wife of Rye Page, owner and publisher of the *Star News* newspaper, she made sure the club received good press and some printing favors.

Jacqueline White was the first Azalea Festival queen. Her most highly acclaimed

Jacqueline White under the Airlie Oak. (Photo by Hugh Morton, Henry B. Rehder Collection)

work was "The Narrow Margin," produced four years later. Though not as well-known at the time as some other stars, she turned out to be the perfect queen for the first festival. Well-traveled, beautiful, and accustomed to protocol because of her family's military background, Jacqueline White fit all the requirements of her position. She seemed equally at ease chatting with Wilmington's most accomplished citizens at the Orton Plantation barbecue, or with tongue-tied schoolchildren at Greenfield Lake.

And she learned quickly that Queens are highly scrutinized. A reporter present at the St. John's Tavern luncheon wrote, "She took no dessert and smoked no cigarette following luncheon." But how refreshing the first meals of

St. John's Tavern as it appeared in 1948.

the day must have seemed: Festival officials saw to it that the Cape Fear Hotel staff served her breakfast in bed during her stay.

At her coronation ball at Lumina, Miss White's composure was tested when Governor R. Gregg Cherry crowned her Queen Azalea I. The first citizen of North Carolina had been in town all day, enjoying seeing a number of old friends. After spending hours socializing, the elderly gentleman was somewhat overdosed on

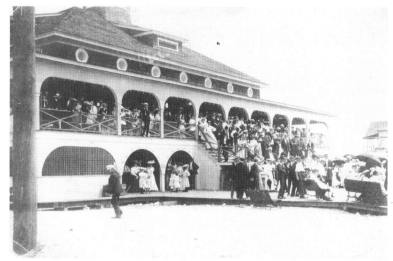

Lumina, pictured here much earlier, was the sight of early Azalea Festival coronations. (Cape Fear Museums)

Southern hospitality. He teetered dangerously close to the edge of the stage before placing the crown upside down on Miss White's head. According to president Hugh Morton, Carl Goerch, editor of *The State* magazine and chairman of the coronation ball, "died a thousand deaths." Miss White merely beamed and the crowd loved it. After the ceremony, the dancing continued until one am.

It was quite an evening for Hugh Morton. He had ushered Dr. Moore's dream into reality and witnessed the crowning event in Lumina, the pavilion created by his grandfather, Hugh MacRae, and named by his mother Agnes.

Jacqueline White Anderson returned in 1972 for the 25th Azalea Festival and shared the stage with her old festival friend, Hugh Morton. She lives on in Wilmington through a much-published photo Mr. Morton took of her under the Airlie Oak.

The Airlie connection conjured up old tour memories for some. In the early 1920s, owner Sarah Jones Walters, widow of Pembroke Jones and wife of Atlantic Coast Line president Henry Walters, opened Airlie to the public for a springtime tour. The donated ticket proceeds helped build and maintain St. Andrews on-the-Sound, the Episcopal church on Airlie Road. The tours continued into the 1930s and sometimes Orton joined in, also contributing its ticket take to the church.

After Mrs. Walters's death, her daughter, Mrs. John Russell Pope, sold Airlie to the W. A. Corbett family of Wilmington. The sale occurred in 1948, the same year the Azalea Festival began, and W. A. Corbett and his descendants have assisted with the festival in some way ever since.

For several years, the Corbett family hosted a Patrons' Party at the 39-room Jones-Walters Mansion (now razed). According to Josephine Corbett Horton, a daughter, "Orton used to have a barbecue on Friday at lunch and Airlie provided a meal on Sunday. The luncheon was in the house as long as the house was there."

The house at Airlie was well-suited for parties. Guests during the Jones-Walters era included presidents of all the major railroads, Newport residents in the famous Gilded Age, and one of the world's wealthiest women, Mary Lily Kenan Flagler. The Corbett family graciously continued the hospitable traditions of Airlie. Even the grand old din-

Festival dining room decor at Airlie, 1948. (Photo by Gilliam K. Horton, Josephine Corbett Horton Collection)

Owner Bertha Barefoot Corbett's siblings, Ola Barefoot Pierce and Dr. Graham Barefoot, Wilmington's first radiologist, attended the Airlie festival parties. The staircase, acquired by former Airlie resident Sarah Jones Walters, once graced the home of Sir Walter Raleigh. (Photo by Gilliam K. Horton, Josephine Corbett Horton Collection)

ing room table was the same: an enormously long mahogany piece that the Corbett family gave eventually to the University of North Carolina at Wilmington. For festival events, the table was set famously with white linen, pink azaleas, and 5-branched silver candelabra holding white tapers.

The Corbett children, particularly Dorothy Corbett Davis and Josephine Corbett Horton, spent long hours helping their mother put on the lavish and elegant event. Also well-known Airlie employees Thelma Mack, Mariah Mandy, Preston Mack, Houston Tartt and W. A. Taylor worked hard to make the Airlie event a seamless affair.

Mrs. Royce McClelland catered the Airlie party one year and Hugh Morton set up headquarters for all the photographers in a "large space that was two bedrooms that opened into one huge room," said Mrs. Horton.

"Eventually Orton quit having the Friday party and then Airlie took the Friday slot."

The Patrons' Party at Airlie continued to be Wilmington's hottest ticket throughout the era of Corbett ownership. In 1999, New Hanover County purchased a 67-acre portion of the tract from the Corbett family. The party goes on, drawing an average of 2,000 people every year. Although it occurs on the last day of the workweek, the dress is anything but "casual Friday" as patrons in their dressy best complement Airlie's vivid spring colors.

Hugh Morton, being a mighty oak of the festival and a grandson of charter garden club member Mrs. Hugh MacRae, has always been a treasured friend to both organizations. The two of them took advantage of one of his most beautiful photographs: Cape Fear Garden Club and the Azalea Festival both used colorized versions of the same Hugh Morton photo of Greenfield Lake in 1948-49. The festival used it as the cover of its streamlined early programs. The garden club used it as a yearbook cover. A few years later, it was Mrs. Hugh Morton who suggested a motto for the garden club: "and He planted a garden and there He put man." (Genesis 2:8)

Dr. Houston Moore saw the first festival but died the following July. However, he left his avocation in the hands of an ingenious assistant. By December 20, Hugh Morton had produced a movie about the first festival to use as publicity for subsequent events. Filmed by Holly Smith of Charlotte, written by Mr. Morton, and narrated by Ted Malone, it debuted in Brogden Hall to an audience of 2500.

That same week, Cape Fear Garden Club hosted a benefit that would, in turn, help fund the upcoming festival's flower show. On December 18th and 19th, the club hosted an elegant Christmas decoration exhibit. The chief participants were Mrs. P. R. Smith, Allie Morris Fechtig, Mrs. Cyrus Hogue, Jane LeGrand, Mrs. Graham Kenan, Mrs. Marsden Bellamy, Mrs. Clark James, and Mrs. Sam Nash.

Martha Hyer in her Coronation Ball gown, 1949. (Photo by Hugh Morton. Cape Fear Museum)

1949

After just one year, the Azalea Festival program grew to include a few more events, including trotting races and an 800-voice community choir concert at Legion Stadium, "Teen Agers Club Azalea Ball," and the popular Azalea Open Invitational Golf Tournament. The tournament existed from 1948 until 1971. According to J. B. Hines, the prize for the first trophy was $10,000 and the last one was $65,000. Golfers like Arnold Palmer and Jack Nicklaus took part in the event, staged on the links of Cape Fear Country Club.

Another popular feature of the 1949 Azalea Festival was a downtown street dance, an event that drew 2000 people. The one hundred block of Chestnut Street was roped off, just as it had been during the Feast of Pirates street dances. Three orchestras played late into the night.

Broadcaster Ted Malone returned to Wilmington, accompanied by his wife and producer. A famous pilot, Bill Odom, who was known for his daredevil flying stunts, brought them to Bluethenthal Airport April 1. Before the year's end, Odom died during the Thompson Trophy Race, an air racing event. A woman and child were also killed, ending sanctioned air racing events nationwide.

Mrs. Hilda Burnett, a special guest, won a trip to the festival because she was the victorious competitor on a Mutual Broadcasting radio show called "Queen for a Day." Her predecessor, Barbara Randall, had visited in 1948. Clothier Beulah Meier provided wardrobes for the women, and the Azalea Festival and Cape Fear Hotel provided them breakfast in bed.

Azalea Festival Queen II was Martha Hyer. Crowned by football star Charlie "Choo-Choo" Justice, she reigned over a late night program at Lumina that was aired by the Tobacco Network Broadcast Company. After the beach event, the entire official party moved along to a private party at Cape Fear Country Club.

An RKO star who would in 1954 star with Audrey Hepburn in the classic film "Sabrina" and be nominated for a 1958 Academy Award for the film "Some Came Running," Martha Hyer counted as friends many actors whose names are everyday vocabulary like Marlene Dietrich, Frank Sinatra, Grace Kelly, William Holden, Tony Curtis, Elizabeth Taylor, and Mia Farrow. In 1990, Martha Hyer completed an autobiography entitled *Finding My Way* in which she recorded her show business experiences and also revealed that in 1949 she depended on $25 a week she "received in government unemployment compensation as a lifeline." But the Azalea Festival bet on a rising star: Martha Hyer's resume would eventually include roles in almost 70 motion pictures.

The memory of the festival is still fresh to Martha Hyer Wallis. "As you know, at the time I was just beginning my film career," she said in 2003, "but as Queen of the Azalea Festival I was a *star!* ...I experienced Southern hospitality at its best at the Festival. I'll always remember the beauty of Wilmington and the heartfelt welcome of its people."

Hannah Wright, Gen. George C. Marshall, Mrs. Marshall, and Bishop Thomas H. Wright pose at Greenfield Lake during the Azalea Festival in 1949. (Photo by Hugh Morton, Cape Fear Museum)

General George C. Marshall was *grand* marshal of the Azalea Festival parade in 1949. Marshall, a five-star general, was Army Chief

Photographer Gilliam Horton labeled this photograph "Festival Day 1949." Mr. Horton's work creates a valuable record of Airlie, shortly after Mr. and Mrs. W. A. Corbett purchased the showplace. (Photo by Gilliam K. Horton, Josephine Corbett Horton Collection)

510 Orange Street. Reserved and self-effacing, the general was a charismatic festival guest. Fifty-four years later, Hugh Morton said that General Marshall was "the greatest Azalea Festival celebrity *ever*."

of Staff, Secretary of Defense, Secretary of State, and the author of the Marshall Plan, a formula to rebuild Europe after World War II. He had come to know Wilmington native Thomas H. Wright when Bishop Wright was a young man serving as chaplain of Virginia Military Institute. The Marshalls were house-guests of the Wrights at the "Bishop's House,"

1950

Actress Gregg Sherwood, Queen Azalea III, reigned the same year that she starred in

Gregg Sherwood. (Photo by Gilliam K. Horton, Josephine Corbett Horton Collection)

Airlie's Spring Garden, Mrs. Henry Walters's favorite garden space, in all its splendor, about 1949. (Photo by Gilliam K. Horton, Josephine Corbett Horton Collection)

"The Golden Gloves Story." As done for the other early queens, her Azalea Festival wardrobe was offered when she was invited to the celebration. Though the queens usually brought along some of their own clothes, most were advised by legendary Wilmington cloth-

W. A. Corbett and Gregg Sherwood at Airlie, 1950. (Photo by Gilliam K. Horton courtesy of Josephine Corbett Horton)

ier Beulah Meier, who also dressed the Queen's Court and other high-profile women in the early years of the festival. Mrs. Meier, who customized clothes to each girl's coloring and shape, traveled to New York to buy fabric, employed dressmakers, oversaw alterations, and added glamour and style to the festival. Beulah Meier was the closest thing Wilmington had to a Coco Chanel.

For the first two years of the festival, Beulah Meier directed the coronation ceremony. In 1950, Ms. Meier dedicated her attention to gowns for the ever-growing number of

celebrities, and Hannah Block and Willa Dickey took charge of the coronation's production. One of their proudest products was a program, "Southern Belles," in which all the women were outfitted in colorful antebellum-style dresses and hats.

Hannah S. Block, a civic-minded Virginia native, directed a total of 25 coronations. She also coached Miss North Carolina contestants, producing one Miss America, Maria Fletcher. Interviewed in 1969 by *Star News* writer Louise Lamica, Ms. Block said, "Mrs. Meier had impeccable taste, and Miss Dickey was ingenious. I would never have been able to stage a pageant through those many years without them....It takes nearly a full year to plan and form a pageant...I have nothing but praise and admiration for both of them. They were troupers all the way."

Queen Azalea III brought her own clothes to Wilmington, but Rye Page learned they were not nearly so conservative as festival guidelines dictated. He called Beulah Meier and asked her to meet Ms. Sherwood in her suite at the Cape Fear Hotel. Mrs. Meier took the clothes back to her shop and added backs and more front to Gregg Sherwood's Hollywood wardrobe.

Gregg Sherwood, Rye Page, and Valera Corbett at Airlie. Gregg Sherwood married Horace Dodge, Jr., and became known as Dora Dodge. (Photo by Gilliam K. Horton, Courtesy of Josephine Corbett Horton)

Airlie was a garden highlight of the early festivals. Here, the old billboard depicts the famous black and white swans and the painted pergola. (All photos on this page were taken by Gilliam K. Horton and are used courtesy of Josephine Corbett Horton.)

Julius Evans, beloved right-hand man to Pembroke Jones, Henry Walters, and W. A. Corbett.

Artist Minnie Evans gained inspiration from Airlie's blooms and sold her artwork at the Airlie gate.

The pergola in silhouette, about 1948.

Minnie Evans and W. A. (Bud) Taylor pose at the Airlie gate. about 1948.

Azalea Festival guests enjoy the lake at Airlie, about 1949.

Claude Howell designed the 1950 cover. (New Hanover Public Library)

By 1950, several garden clubs had sprouted from Cape Fear Garden Club seeds. Sometimes they duplicated one another in fundraisers and exhibitions, so Cape Fear Garden Club created the Garden Council of New Hanover County. The first meeting took place in the Great Hall of St. James Episcopal Parish House on January 16, 1950. The new program of coordinated efforts and some shared projects created harmony and produced some synergistic campaigns. As a result, garden clubs proliferated and membership waiting lists elongated in New Hanover County.

Azalea beauties were occasionally photographed at Belk-Beery in the 1950s. Located in the same building that now houses the New Hanover County Public Library at Second and Chestnut streets, Belks was downtown's leading store when downtown was virtually the only place to shop. This particular group of women features Faye Arnold (Broyhill) who was Miss N. C. in 1955 and 3rd runner up for Miss America in 1956. (Cape Fear Museum)

1951

Margaret Sheridan's stint as Queen Azalea IV practically coincided with her role in "The Thing," director Howard Hawks's sci-

Robert Bellamy, Josephine Corbett Horton, Patsy O'Shea, and Tom Brown at the Airlie Azalea Festival party, about 1950 (Photo by Gilliam K. Horton, Josephine Corbett Horton Collection)

ence-fiction classic. The year 1951 proved to be a good one for planning. Public relations work that year led *Life* magazine to include some of Wilmington's gardens in its March 1952 issue.

As the parade continued to grow, new bands were added. But one band always had the most applause along the way, and brought the entire reviewing stand to its feet: The Williston Marching Band.

Williston Marching Band. (Photo by Hugh Morton, Cape Fear Museum)

Also the Cape Fear Garden Council began planning a wrought-iron arch at Greenfield Lake. Dedicated by Mayor E. L. White on March 28, 1952, the arch was designed and built by Vernon W. Toler. The City of Wilmington provided the foundation and the stonemason was Robert H. Brady, Jr., who used marble from the Carolina Cut Stone Company.

1952

According to Hugh Morton, actress Janet Leigh was scheduled to reign over the

Bill Sharpe (on left), Tar Heel publicist, and Kenneth and Betsy Sprunt (on far right) enjoy a festival picnic, about 1952. (*Star News* photo)

Azalea Festival in 1952, but her husband, Tony Curtis, cancelled her engagement despite pleas from Hugh Morton, who had traveled to New York to cement the agreement. Mr. Morton was aware that actress Cathy Downs was in town with her husband, Joe Kirkwood, Jr., an actor who would eventually play the title part in "The Joe Palooka Story" and a golfer who was taking part in the Azalea Open. Morton asked Cathy Downs if she could step into Ms. Leigh's place. She accepted and the Azalea Festival committee flew her to Fayetteville so that she could arrive with fanfare on a publicized Piedmont Airlines flight. Cathy Downs reigned over the 1952 festival, and enjoyed it so much she returned the next

Cathy Downs (Photo by Gilliam K. Horton courtesy of Josephine Corbett)

year as a "commoner." Among other movies, Cathy Downs was in "My Darling Clementine" in 1946, and would star with her husband in 1952 in "The Joe Palooka Story." The 1952 festival featured a street dance at 17th and Market, parties at Lumina, and a production of "H.M.S. Pinafore" at Legion Stadium.

During those early years of the festival, members of Cape Fear Garden Club were learning that they could create garden shows that delighted, instructed, and made a profit that could be recycled into additional beautification. Notable garden club events included assorted flower shows, camellia displays, the annual Christmas House,

Claude Howell's Spring Garden Show cover, 1952. (Louise Wells Cameron Art Museum)

"...toppers" topped with flowers constituted the end of the track in Wilmington for decades. During the early ...ars, thousands of festival guests arrived at Union Station. (Fales Collection, Lower Cape Fear Historical ...ociety)

project, featured plants and shrubs that would attract birds for the patients to see.

Assorted "worshipful" projects included funds to Little Chapel on the Boardwalk, a "bulb shower" to St. James Church, and trees to B'Nai Israel Synagogue and Bethany Presbyterian Church. Additionally, donations were made to the Tryon Palace Restoration Fund, both for the building and to acquire furnishings.

But each year, Cape Fear Garden Club was merely participating in the spring Pilgrimage. It was a lot of work and had become a kitchen with many cooks. Bess Smith made some careful observations and, by the next year, the garden club reigned supreme. Not only did they have Henry Rehder's blessing, they also had him. From the first Pilgrimage until 2003, Mr. Rehder has been involved in the tours, often as a garden owner.

and an especially successful Spring Flower Show, held at the Surf Club in 1952. The Wrightsville Beach extravaganza, chaired by Mrs. Emmett H. Bellamy and Mrs. Graham Barefoot, brought rave reviews and several state awards.

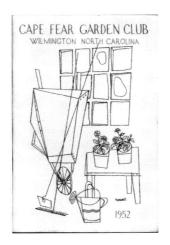

Frozen animation — another cover by Claude Howell. (New Hanover County Public Library)

Proceeds from the shows were channeled into a variety of causes, both in town and "away." Members voted to give donations of cash or decorations to the Azalea Festival Committee, Camp Lejeune Hospital, the Catherine Kennedy Home, the American Red Cross, a garden behind the downtown post office, the New Hanover County Home, and the John C. Wessell Tuberculosis Sanitorium, off Castle Hayne Road. The sanitorium, a pet

Bert Parks and his wife were house-guests of Barbara Beeland Rehder (on right) and her husband, Henry, in 1956. Mr. Parks was most famous for singing, "Here She Is, Miss America," on the annual pageant telecast. But he won Wilmington's heart by visiting the sick in Bulluck Hospital. (Henry B. Rehder Collection)

35

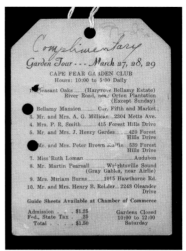

An original Cape Fear Garden Club Azalea Tour ticket, 1953.

The Cape Fear Garden Club Tours

1953

Mr. and Mrs. Hargrove Bellamy, Pleasant Oaks, River Road, in Brunswick County

Bellamy Mansion (1859), corner 5th Avenue and Market Street, with first floor of mansion open

The Bellamy Mansion. (Photo by Barbara Marcroft, Cape Fear Museum)

Mr. and Mrs. A. G. Millican, 2304 Metts Avenue

Mrs. P. R. Smith, 615 Forest Hills Drive

Mr. and Mrs. J. Henry Gerdes, 722 Forest Hills Drive

Mr. and Mrs. Peter Browne Ruffin, 753 Forest Hills Drive

Mrs. Ruth Loman, 206 Audubon Boulevard

Mr. Martin Pearsall, Wrightsville Sound on Airlie Road (now known as Gray Gables)

Mrs. Miriam Burns, 1417 Hawthorne Road

Mr. and Mrs. Henry B. Rehder, 2217 Oleander Drive.

Cape Fear Garden club members were jubilant to present their first tour. When they took over the old Pilgrimage Garden Group Tour, they renamed it the Spring Garden Pilgrimage. The motion to run the tour was made by Mrs. P. R. Smith in a garden club meeting in her garage, at 615 Forest Hills Drive. Mrs. Smith, who had helped run the family business, Cape Fear Ford, was methodical in examining the working of the garden club. It was she who requested, in 1951, that all records of the club be preserved, compiled, and donated to the Wilmington Public Library.

"Bess Smith was the great organizer of the club at that time," said Henry B. Rehder. Mrs. Smith made sure that the garden club, standing alone, could run the tour efficiently and appropriate the proceeds wisely. Club members set $1.50 as the ticket price and agreed to give one third of the profit back to the Azalea Festival. Gardens would be open daily throughout the festival, with the exception of Saturday morning during the parade. A modest honorarium would be offered to garden owners.

Tour planning was time-consuming and club members stayed especially busy that winter and spring. "With less than a month to go," stated the *Wilmington Morning Star* on February 28, 1953, "everyone is working now completing plans for the Azalea Festival. Cape

Fear Garden Club is planning a tour of a number of our most beautiful private gardens which will be open during the Festival."

The tour took place March 27-29 and the officers included: Mrs. Charles Cavenaugh, chairman; Mrs. J. W. Grise, chairman of hostesses; and Mrs. J. J. Burney, Jr., ticket chairman. The ten gardens on that first tour provided guests with a cornucopia of Cape Fear's greatest natural resources and rich displays of gardening expertise.

The Gerdes garden at 722 Forest Hills Drive featured several transplanted items of interest. A banana shrub from a Charleston plantation, fig vine from Brookgreen Gardens, and a tree magnolia grown from a cutting taken from the Burgwin-Wright House garden added interest to the garden. Local cobblestones were also incorporated into the landscape design.

Along with the tour, parade, and concerts, there were lots of festival parties by 1953, particularly for young people. Azalea Queen VI Alexis Smith made an appearance at many of them. A 12-year veteran of Warner Brothers productions, her name was widely known. She continued her show business career into the

1990s, winning a Tony award in the early '70s, and having a role in the movies "Tough Guys" (1986) and "The Age of Innocence" (1993). Ms. Smith was a good sport, too: Another queen had been chosen but became ill just before boarding her flight to Wilmington. Alexis Smith came on four hours' notice.

Arthur Smith and the Crackerjacks entertained local country music fans. In 1951, Smith began filming a regular television show that would last for 32 years. A guitarist and composer, his most famous creations were "Guitar Boogie" and "Dueling Banjos," the song from the movie, "Deliverance." Writer Ralph Grizzle suggested that "Dueling Banjos" might have been inspired by Arthur Smith's effective defense of Azalea Festival icon Hugh Morton, when the government threatened to cut the Blue Ridge Parkway straight through Morton's property, Grandfather Mountain. Smith debated a representative of the engineers for the National Park Service on WRAL-TV in Raleigh and won overwhelming support for Mr. Morton when he said, "When a man like Hugh Morton loves a mountain like he loves Grandfather mountain, it don't seem right for a big bureaucrat to come down here from Washington and try to take it away from him." The result was a cantilevered highway, an almost miraculous engineering feat.

The 1953 Teen-Age Azalea Princess was Mary Frances Davis. Her attendants from various N. C. cities were escorted by New Hanover High School students: Sonny Jurgensen, Jere Partrick, Jon Gerdes, Jimmy Post, Warwick Porter, and Richard Cox. Azalea Festival teenage division parties

Alexis Smith (Photo by Hugh Morton)

The floral poetry of Orton Plantation. (Photo by Hugh Morton)

were held at Lumina, Josephine Debnam's Wrightsville Beach home, The Ark restaurant, and Crystal Restaurant.

Most of the Azalea Pilgrimage gardens were opened again the following weekend, April 2-5, for the 5th Annual Azalea Open Invitational Golf Tournament. Local amateurs included: Robert M. Williams, Jr., J. Holmes Davis, Jr., and Lawrence Cook. Jerry Barber, Cary Middlecoff, Bob Toski, Art Wall, Jr., Mike Souchak, Doug Ford, and Tommy Belt were some of the professional competitors.

On June 19, 1953, the Cape Fear Garden Club executive council met and declared the First Annual Azalea Pilgrimage Tour a blooming success. The proceeds amounted to $600. The ladies voted to make it an annual event.

Also in 1953, groups of ladies from Cape Fear Garden Club drove to Roanoke Island, at Manteo, to witness the groundbreaking for the 10 1/2-acre Elizabethan garden. The brainchild of Sir Evelyn Wrench, founder of the English-Speaking Union, the garden's founding was spearheaded by Edenton's resident author, Inglis Fletcher, and Mrs. Charles Cannon of Concord. Cape Fear Garden Club gave gener-

ously to the project, created as a memorial to the colonists of Roanoke Island. Construction began June 2, 1953, on the elaborate space, the day of Elizabeth II's coronation. In 1984, Britain's Princess Anne visited the Elizabethan Garden for the 400th anniversary of the first English settlement.

1954

Mrs. Jessie Kenan Wise, 1709 Market Street

Mrs. Julia Seigler, 106 Borden Avenue

Mr. and Mrs. Ralph Bertram Williams, 708 Forest Hills Drive

Mrs. Bruce B. Cameron, Sr., 726 Forest Hills Drive

Mr. and Mrs. Arthur E. Tousignant, 2716 Columbia Avenue

Mr. and Mrs. Donald MacRae Parsley, Live Oaks

Mrs. Edwin A. Harriss, Tremont at Eshcol, Masonboro Sound

Mr. and Mrs. A. G. Millican, 2304 Metts Avenue

Mrs. P. R. Smith, 615 Forest Hills Drive

Mr. and Mrs. Peter Browne Ruffin, 753 Forest Hills Drive

Bare floats reflect chilly weather. A northeaster visited during the 1954 festival. (Lower Cape Fear Historical Society)

By 1954, the festival, in the words of Hugh Morton, had gone from a "puny infant to a strapping giant." As in the days leading to Christmas, the newspaper carried a countdown for weeks before the festival. Events filled almost every waking hour and kept many wide-eyed when they were usually sleeping. In the midst of the seventh festival, one hearty partygoer quipped, "It's grand to have the Festival. But isn't it grand it doesn't come but once a year."

The 1954 tour, chaired by Mrs. J. W. Grise, took place March 26-28. Mrs Edward B. Ward was ticket chairman that year, a convenient situation of sorts, since Mr. Ward was Azalea Festival president. Weather threw their organized efforts a curve: on festival Saturday, the high temperature was 60, with a northeast wind. And though a recent freeze had already destroyed the blooms, the tour was a popular one.

The Cottage Lane Art Show that began in 1953 as the Sidewalk Art Show, expanded the second year. Held in the little lane adjacent to First Presbyterian Church, its location had special significance to the local art community.

At one time the lane had contained more cottages, one of them particularly important to Wilmington art enthusiasts: Elisabeth Chant had lived there. Ms. Chant created memorable paintings and also taught notable local artists like Claude Howell.

Jimmy McKoy, a nephew of Lincoln Memorial architect Henry Bacon, thought up the idea of a festival art show and presented his idea to some friends at St. John's Tavern. Gar Faulkner and Claude Howell organized the first art show and Margaret Davis changed the original name to the Cottage Lane Art Show. Mr. Faulkner suggested that the 1954 show be modeled after Parisian shows he had attended. According to a *Star News* reporter, "Members spent months working on paper mache masks

On Cottage Lane, artist Claude Howell cavorts in costume for Queen Ella Raines during the 1954 festival. In those days, the show featured local art and photography. Lifelong travelers Gar Faulkner and Mr. Howell planned the event. According to the 1955 brochure, it was "an outdoor exhibit, styled to Paris tradition." (Cape Fear Museum)

with added decorations like eggshell eyes and spool noses."

Wilmingtonians Elizabeth McEachern and Leila Henderson, known during festival season as the Mad Flatters, lined up entertainers who performed at the art show from the hour it opened until the hour it closed. The Mad Flatters also encouraged a prominent but small group of Wilmingtonians to wear costumes to a private Front Street festival party that evolved into an annual event. One year, the irrepressible Emma Bellamy Williamson Hendren, arrived wearing a tutu.

Queen Ella Raines. (Photo by Hugh Morton, Cape Fear Museum)

Mrs. Mark Clark, wife of The Citadel president, cut the ceremonial ribbon and the general crowned Queen Ella Raines, who played the heroine in the 1943 film "Phantom Lady" and starred in the television series. The 1954 model crown was a doozie. Loaned by Harry Winston, Inc., of New York, the tiara contained 1900 diamonds and 75 rubies and was once owned by Franz Josef, 19th-century Emperor of Austria and King of Hungary. The diamonds alone weighed more than 180 carats.

The sight of the $750,000 treasure, sparkling in the lights of Lumina, brought gasps from the audience.

Charlie "Choo-Choo" Justice returned for the 1954 festival. The football player, who in his time was as popular in North Carolina as Michael Jordan would be in his, was mobbed almost everywhere he went. Anita Colby, an actress and beauty consultant, was also a popular celebrity.

What the gardens lacked in blooms that year, they made up with the addition of some grand new showplaces. The opening of the Wise Garden in 1954 presented a particularly attractive opportunity. The property belonged to Jessie Kenan Wise (1870-1968), a garden club member and a sister of Mary Lily Kenan Flagler (1867-1917), one of the world's most famously wealthy women.

Sarah Graham Kenan stands in front of the Wise House. (Photo by Thomas S. Kenan III)

Wilmingtonians who had had only a shadowy glimpse of Mrs. Wise through the

A white marble fountain in the Kenan Garden. (Photo by Thomas S. Kenan III)

back window of her chauffeured limousine, could now tarry in her garden and closely study the carefully manicured grounds. Some talked of the time a local man stopped to help Mrs. Wise with a flat tire. The next day he received an unexpected gift: a new truck from the grateful dowager.

The Kenan House, 1956. (Photo by Thomas S. Kenan III)

Another sister, Sarah Graham Kenan (1876-1968), a member of Cape Fear Garden Club, owned the Kenan House, now next door to the Wise House. Both sisters willed their Wilmington houses to the University of North Carolina at Wilmington. The Kenan House serves as residence to the Chancellor. Mrs. Wise's home, now restored, is known as the Alumni House.

On June 2, 1954, a committee of three garden club members chose Marie Rehder Gerdes to write the official Cape Fear Garden Club Collect. The result was a prayer that was

The Doll House at Airlie, built by Sarah Jones for her granddaughters, was a perennial hit for garden tourists. The Corbett children enjoyed the Airlie doll house until Hurricane Hazel destroyed it, in 1954. The Airlie Mansion survived Hazel but was razed to make way for new construction. (Photos by Gilliam K. Horton, Courtesy of Josephine Corbett Horton)

adopted by the Garden Club of North Carolina, Inc., the following August, and has been reprinted in periodicals ever since. Ms. Gerdes's Collect reflects her decades-long passion for gardening and Sunday School teaching, her preference for the King James translation of the Bible, and her belief that gardeners never really work alone.

"Our Heavenly Father, Who dost feed the birds and clothe the flowers, and Who knoweth and careth for every need of us Thy children so enlighten our minds to use wisely all the gifts of Thy Mighty Hand that we, being imbued with Thy Holy Spirit, may so work Thy will that those who come after may mark their path by our footsteps.

"For all the beauty of the earth, Father in Heaven, we thank Thee. For our families, our friends, our free and beautiful country, Father in Heaven, we thank Thee.

"We beseech Thee of Thy great goodness and tender mercy to forgive our sins and grant that as we work together in fellowship we may draw closer to Thee, Almighty God, in whose name we pray. Amen."

An Airlie delicacy. (Photo by Gilliam K. Horton, courtesy of Josephine Corbett Horton)

Miriam Burns's garden was named Garden of the Year by the state club in 1955. A Juilliard graduate, she built it around a musical theme. (*Star News* photograph, Miriam Burns Whitford)

1955

Mrs. Miriam Burns, 1417 Hawthorne Drive

Mr. L. F. Duvall, 214 Forest Hills Drive

Judge and Mrs. John J. Burney, 1516 Chestnut Street

Atlantic Coast Line Railroad, 119 Red Cross Street

Martin Pearsall, Airlie Road

Mr. and Mrs. Henry B. Rehder, 2217 Oleander Drive

Mrs. P. R. Smith, 615 Forest Hills Drive

Mr. and Mrs. Peter Browne Ruffin, 753 Forest Hills Drive

Mr. and Mrs. A. G. Millican, 2304 Metts Avenue

Pleasant Oaks, River Road in Brunswick
County

Mrs. Ruth Loman, 206 Audubon Boulevard

Orton owner J. Laurence Sprunt entertains actress
Ann Sheridan. (Orton Collection)

Mrs. Lewis Ormond was chairman of
the third garden club tour, and she had her
hands full. It began on April 1, an appropriate
date since nature had played a naughty joke on
the azaleas. A recent hard freeze had left the
flowers looking like prunes instead of plums.
The tour, unofficially canceled, still brought
guests, and garden club members did their part
to promote the club and the festival by handing
out greenhouse azalea bouquets.

Though plenty of blooms were there,
they weren't necessary to entice visitors to

Martin Pearsall's garden on Airlie Road. A
huge live oak tree delighted visitors.
Measuring 20 1/2 feet in circumference, the
tree was substantially larger than the Hilton
live oak billed as the World's Largest Living
Christmas Tree, that measured only fifteen feet
around its base. The Pearsall tree's spread of
branches reached a whopping 202 feet. Martin
Pearsall, a close friend of Pembroke Jones, was
a frequent guest at the Bungalow, located on
what is now Great Oaks Drive in the Landfall
subdivision. Mr. Pearsall attended many of the
private festival parties. Rubbing shoulders with
celebrities was nothing new to him: Some of
the early silent movie stars were also guests at
Pembroke Jones's hideaway.

It was a "white gloves day" when Queen Sara
Shane and her court visited Orton. (Photo by
Hugh Morton, Cape Fear Museum)

The eighth festival's top celebrity was
Queen Sara Shane, a successful television and
cinema actress whose favorite movie was
"Tarzan's Greatest Adventure," with Anthony
Quayle and Sean Connery. It was also the

favorite Tarzan of film critics, who marveled that Sara performed all her own stunts.

Beautiful and intelligent, Ms. Shane probably saw some humor in choosing a pretty stage name that sounded like one. Later, she made a career change and took back her original first name and surname by marriage. She is now known as Elaine Hollingsworth, an author, lecturer, and the director of the Hippocrates Health Center of Australia. Her book, *Take Control of Your Health and Escape the Sickness Industry*, has sold 50,000 copies worldwide, with the promise of more sales as many countries are requesting translation and publishing rights. In addition to being an author and lecturer, she maintains her own website, known as doctorsaredangerous.com.

Carl "Chick" Mathis was Sara Shane's escort. Mr. Mathis, single, debonair, and good-looking, escorted visiting festival beauties for ten years. His festival "job" included keeping close company with many Azalea Queens, a Miss America, and one Miss North Carolina. Mr. Mathis, who learned that some queens have a little too much princess in them, remembers Ms. Hollingsworth as a wonderful person and a real trouper. "She performed her festival duties beautifully," he said, "despite the fact that she was hospitalized the day before in Washington, D. C., for an unusually severe case of poison ivy." Wilmington had only two dermatologists at the time, both beloved. While playing golf at Cape Fear Country Club, Dr. Harry Van Velsor got a message to go treat the queen.

But he didn't respond. "I was having a good game. Doctors don't get off the golf course on a weekday if they're playing well — even for the azalea queen. I told them Bill Phillips was on duty."

Dr. William Phillips treated Ms. Shane and advised her against scratching the exasperating itch. But foreshadowing her second career, Sara Shane chose to be her own physician. She told a newspaper reporter that she continued to scratch the rash, hoping it would release the poison from her body: "Doctors won't agree with me on this type of treatment, but it works."

The 1955 festival, especially the 2,000 people who came to the airport to greet her, is still a fresh memory to its queen. "When I arrived in Wilmington, all I wanted to do was go to the hotel and crash," Ms. Hollingsworth reminisced, in 2003. "When I saw all those people outside the plane I thought some important person had arrived. Then when I realized they were there to greet me, I went into shock.

"But somehow, in spite of it all, I had a great time. I remember my time in your lovely city with such fondness and my escort was a real sweetie."

Azalea Festival volunteer extraordinaire Carl "Chick" Mathis says a touching goodbye to Queen Sara Shane, in 1955. Almost fifty years later, they both have their copies of the print. (JoAnne Mathis)

Poet Ogden Nash, a cousin of Mrs. J. Laurence Sprunt, wrote the following poem in the guest book at Orton:

"There was a man,
A social failure—
He said Camellia,
Not Camaylia."

(Photo by Hugh Morton)

Two other beautiful actresses were guests of the 1955 festival: Kim Hunter and Polly Bergen. Ms. Hunter, who won an Academy Award for her portrayal of Stella in "A Streetcar Named Desire," would go on to appear in many more movies, including "Requiem for a Heavyweight," "Planet of the Apes," and "Midnight in the Garden of Good and Evil." Polly Bergen would return the following year as Queen Azalea X. Both actresses were stars of the parade, an event that was spawning more and more galleries in unusual places. The *Star News* reported that at St. James Church, "members of the congregation were sitting out in front of the building before 8 a.m. Benches from the Sunday School served as seats for the early risers."

The ribbon cutting, 1956. In the foreground: Miriam Burns (in white hat) and son, Jim (in bow tie). (Mimi Burns Whitford)

1956

Mrs. Miriam Burns, 1417 Hawthorne Road

Dr. and Mrs. Graham Barefoot, 206 Forest Hills Drive

Mr. and Mrs. Peter Browne Ruffin, 753 Forest Hills Drive

Mrs. Bruce B. Cameron, 726 Forest Hills Drive

Mr. and Mrs. Ralph Bertram Williams, 708 Forest Hills Drive

Mrs. P. R. Smith, 615 Forest Hills Drive

Mr. and Mrs. A. G. Millican, 2304 Metts Avenue

Mrs. Edwin A. Harriss, Tremont at Eshcol, Masonboro Sound

Mr. Martin Pearsall, Airlie Road

Mr. and Mrs. Henry B. Rehder, 2217 Oleander Drive

Atlantic Coast Line Railroad Garden, 119 Red Cross Street

Tree projects took precedence with the garden club in 1956. Members celebrated the 100th anniversary of President Woodrow Wilson's birth by dedicating a handsome marker at First Presbyterian Church, noting a crepe myrtle tree planted there by his mother, Janet Woodrow Wilson, in 1880. The future president's father was minister of First Presbyterian Church from 1874 to 1882. Though much of that time he was away at school, he was here long enough to make some lifelong friends, including, among others, James Sprunt, Hugh MacRae, John Allan Taylor, Benjamin Franklin Hall, John D. Bellamy, Jr., and Henry C. McQueen. In the 1950s, there were still people here to retell the story of the day young Woodrow misjudged the slope at the foot of Nun Street and rode his celebrated bicycle into the river.

There were many garden projects that

It has been said that Woodrow Wilson had the first bicycle in Wilmington. This sketch may be the work Kenneth M. Murchison. (Special Collections, Duke University)

year. Mrs. Henry E. Longley supervised the planting of 138 dogwood trees at the Air Force Radar Station at Fort Fisher. Mrs. John J. Burney and Mrs. J. Henry Gerdes, who had recently gained national status as flower show judges, were busy helping others who were studying for accreditation: Mrs. Graham Barefoot, Mrs. Emmett H. Bellamy, Mrs. Leslie N. Boney, Mrs. R. T. Davis, and Miss Monimia MacRae.

Mary Lily Boney, a member of Cape Fear Garden Club and an artist who painted flowers, poses with her beloved orchids at 120 South Fifth Avenue. (Courtesy of William J. Boney, Jr.)

The Atlantic Coast Line Railroad Garden made another appearance on the tour roster. A pet project of Champion Davis, it was meticulously laid out on three levels, with superior varieties of trees, azaleas, roses, pansies, tulips, camellias, and shrubs. Walkways of pink marble added even more luster and color to the brilliant garden. It was a welcome discovery for tourists and a treasured familiar space for those who worked at the Atlantic Coast Line. The Coast Line Garden looks different now, but it still exists. Eventually Mr. Davis gave it to the city of Wilmington, a precious green space for the public's pleasure.

Prior to the 1956 ribbon cutting, Cape

Polly Bergen, flanked beautifully by Ann Hart and Janet Rudolph of Cypress Gardens, poses at Airlie. "Dick Pope of Cypress Gardens always made sure we had some fine girls," said Catherine Meier Cameron. (Photo by Gilliam K. Horton, Josephine Corbett Horton Collection)

Fear Garden Club members, directed by Mrs. Eugene Ballard, hosted a coffee hour for queen Polly Bergen at Cape Fear Country Club. The queen's court and the governor were also special guests. Then the official party traveled a few blocks to the home of Mrs. Miriam Burns, where Ms. Bergen opened the 1956 garden tour.

Flowers were at their peak that year and even a few April showers didn't hurt the tour's success. Chairman Mrs. W. S. Baker and other festival officials all knew that wherever Polly Bergen was, a Pepsi had to be there, too. In fact, at the queen's party in the Burns garden, Polly Bergen arrived with a wheelbarrow full of Pepsis on ice.Official spokesperson for the drink, she was showing her business acumen at age 25. Eventually Polly Bergen would consid-

er herself a businesswoman first and an actress, second. She sold her own line of shoes, jewelry, and cosmetics.

Polly Bergen, whose birth name, Nellie Burgin, is reminiscent of an old Wilmington surname, had roles in many movies and television shows. Ironically, the one best known today is the thriller "Cape Fear" (1956), in which she starred with Gregory Peck and Robert Mitchum.

Ms. Bergen, an A-list star, was invited to the festival by William Broadfoot. "Billy Broadfoot got some of the early queens," said Chick Mathis. "He spent all year on it and would go to California to find the best queen. He was a great salesman and the best recruiter."

According to Hugh Morton, the festival

47

originally contacted over a dozen movie studios, to be welcomed only by RKO. Festival organizers worked through a helpful representative named Lynn Unkefer.

Lottie Fales Cameron in her garden at 726 Forest Hills Drive. (Hilda Cameron Echols)

1957

Mrs. Bruce B. Cameron, 726 Forest Hills Drive

Mr. and Mrs. J. Kyle Bannerman, 2502 Market Street

Mrs. William G. Broadfoot, Sr., 148 Colonial Drive

Mr. and Mrs. Hugh MacRae II, 1212 South Live Oak Parkway

Col. and Mrs. George W. Gillette, 1236 South Live Oak Parkway

Mr. and Mrs. S. L. Marbury, 741 Forest Hills Drive

Mr. and Mrs. Peter Browne Ruffin, 753 Forest Hills Drive

Dr. and Mrs. Graham Barefoot, 206 Forest Hills Drive

Mrs. P. R. Smith, 615 Forest Hills Drive

Mr. and Mrs. Ralph Bertram Williams, 708 Forest Hills Drive

Mr. and Mrs. Hargrove Bellamy, Pleasant Oaks Plantation, River Road in Brunswick County

Pleasant Oaks Plantation, now the home of Lyell Bellamy and Brian McMerty. (Lower Cape Fear Historical Society)

The 1957 Azalea Festival favored Forest Hills Drive. Cape Fear Garden Club assisted Mrs. P. R. Smith in hosting a barbecue in her Forest Hills garden to welcome Queen Azalea X. Her neighbor, Mr. S. L. Marbury, who was president of the American Camellia Society, took part in the festival. And the garden tour ribbon cutting took place at the home of Mrs. Bruce B. Cameron, who lived midway between them. Garden Tour Chairman Mrs. Rufus LeGrand donated the year's proceeds to the Girls Club of Wilmington.

The Kyle Bannerman house, now the residence of Dr. and Mrs. John Cashman, boasted beautiful camellias and large azaleas. Mrs. Bannerman, an active member of Cape Fear Garden Club, also oversaw the landscaping of Post Office Park, a club project.

Many of the 1957 celebrities were well-known nationally, and the events were more crowded than ever. Charles Craven, a writer for Raleigh's *News and Observer*, overheard the following in Brogden Hall during the Coronation: "Lady, you may have my seat if you are actually pregnant."

Out at Cape Fear Country Club, Arnold Palmer dazzled his army when he won the 1957 Azalea Open. But, onstage, comedian George Jessel didn't win points when he announced that the Cape Fear Hotel was so old that George Washington's initials were carved in the men's room. But Dale Robertson, star of the popular "Wells Fargo" television show, delighted the crowds. Queen Kathryn Grayson, a Tarheel, had major movie credits under her little belt. She starred in "Showboat" in 1951 and "Kiss Me, Kate" in 1953. Ms. Grayson's later work included several appearances on the television show "Murder, She Wrote." Opera star Lauritz Melchior crowned Ms. Grayson.

The year 1957 was also the beginning of a Wilmington tradition: Dan and Bruce Cameron's Azalea Festival breakfast

Atlantic Coast Line Railroad Queen Shirlejo Keever (center) is surrounded by members of her court: Betty Tienken, Phyllis Meier, Helen McNeil, and Grace Hobbs. This photo was taken in the ACL garden. (*Star News* photograph, Cape Fear Museum)

at the Cape Fear Club. The annual event became the most sought-after invitation of the season. Movers, shakers, and deal makers mingled with movie stars and television icons while waiters served limitless trays of Bloody Marys. Mrs. Dan Cameron remembers the phone calls that always came before the popu-

A Star News photographer caught the ribbon mid-air when Kathryn Grayson snipped it at the home of Mrs. Bruce B. Cameron in 1957. (Henry B. Rehder Collection)

Mrs. K. M. Sprunt, Mrs. G. K. Horton, and Mrs. J. E. Holton, Jr., at the Burgwin Wright House, take part in a garden club fundraiser. (New Hanover County Public Library)

lar party. "They would ask, 'Can't we just bring along a few friends?'" said Betty Henderson Cameron, in 2003. "And I would say, 'We're already past the fire limit and we have friends of our own we had to leave off the list.'"

1958

Miss Allie Morris Fechtig, 132 West Renovah Circle

Mr. and Mrs. Odell Bridgers, 527 Colonial Drive

Mrs. Miriam Burns, 1417 Hawthorne Road

Mrs. P. R. Smith, 615 Forest Hills Drive

Mr. and Mrs. Ralph Bertram Williams, 708 Forest Hills Drive

Mr. and Mrs. Peter Browne Ruffin, 753 Forest Hills Drive

The 1958 festival was another reminder that nature doesn't follow our calendars. Almost no flowers were in bloom. Since at that time the tour hinged more heavily on azalea blooms, tour chairman Mrs. J. W. Lamberson quickly changed labels, calling it a "Courtesy Tour." Festival official, Cape Fear Garden Club leader, and Greenfield patron Allie Morris Fechtig was especially pleased with a new addition to Greenfield Lake. J. Melville Broughton, director of the N. C. Highway Commission, spoke at the dedication of a new footbridge, but Miss Fechtig cut the ribbon.

In addition to the usual beauty queens, plantation princesses were also selected to represent Orton, Airlie, and Pleasant Oaks. Zeme North was the Orton princess; Toby Keller, a New York model, was the Airlie princess; and the Pleasant Oaks beauty was Paula Lamont.

As a prelude to Azalea Belle attire, Beulah Meier created gowns for the princesses, based on dresses she had seen worn at Cypress Gardens events. Using sample wedding gowns from New York, her daughter, Catherine Meier (Cameron) dyed the gowns in a bathtub to give them color. Then Mrs. Meier crafted flowers from fabric to decorate the dresses. All the flowers and beading were done by hand.

North Carolina native Andy Griffith stayed at the home of his old friend Hugh Morton while in

John Bromfield, Esther Williams, and Andy Griffith at the 1958 coronation. Though Griffith's career hadn't yet "moved" to Mayberry, he had already made a name for himself in N. C. His good friend Hugh Morton encouraged him to take part in the festival. (Photo by Hugh Morton, Cape Fear Museum)

This photo, taken March 28, 1958, shows off the dress Beulah Meier created for Pleasant Oaks Plantation Princess Paula Lamont (on left). Others pictured are (left to right) Scott Brady, King of Hospitality; Elaine Herndon, Miss North Carolina; John Bromfield, King of Festivities, and Henry Rehder, Jr. The backdrop was created by artist Claude Howell. (Henry B. Rehder Collection)

town for the festival. At the time, Griffith was best known for his movie role in "No Time for Sergeants" and his best-selling record, "What it Was, Was Football." Ralph Story, host of the now notorious television show "The $64,000 Question," attracted a lot of attention as well; he served in World War II as a fighter pilot.

Esther Williams, famous for cinematic swimming extravaganzas, reigned over the festival and proved to be a rather feisty queen who mixed festival promotion with large newspaper ads for her swimming pool company.

For a time, she shared the glory with two festival kings: Scott Brady, King of Hospitality, and John Bromfield, King of Festivities. However the event ended early for Mr. Brady. After a disagreement with Esther Williams, some of which happened on stage Friday night, he left before the festival was over. Aviation enthusiast Joe Bennett was at the airport when Brady departed. "I'm not staying here another minute," Brady told Mr. Bennett, as he boarded the plane.

And Ms. Williams's parade gown creat-

A smiling Henry B. Rehder holds Esther Williams's train as she makes her way to the Queen's Float, in 1958. After the parade, Beulah Meier, insisting the festival was a "family event," altered Ms. Williams's dress for the Coronation. Ed Lilly (on right) was a festival supporter and bank executive. (Henry B. Rehder Collection)

ed quite a splash, too. Perhaps spending so much time wearing bathing suits gave her a more relaxed view on necklines than most Azalea Queens. Whatever the cause, Mrs. Williams's parade gown dipped below the threshold of official festival family values, and clothier Beulah Meier was concerned enough to be waiting at the end of the parade route. "I'd like to see you at my shop in one hour," she said. The queen complied and her wardrobe, stored at O'Crowley's Cleaners, was examined for decency. When the alterations were complete, the coronation gown had grown another row of fabric.

Adaptive use on Parade Day. (Photo by Hugh Morton, Cape Fear Museum)

1959

Mr. and Mrs. Henry B. Rehder, 2217 Oleander Drive

Mrs. Miriam Burns, 1417 Hawthorne Road

Mr. and Mrs. Peter Browne Ruffin, 753 Forest Hills Drive

Mr. S. L. Marbury, 741 Forest Hills Drive

Mr. and Mrs. J. Henry Gerdes, 722 Forest Hills Drive

Mr. and Mrs. Ralph Bertram Williams, 708 Forest Hills Drive

Mr. and Mrs. William G. Broadfoot, Jr., 226 Colonial Drive

Mr. and Mrs. A. G. Millican, 2304 Metts Avenue

Atlantic Coast Line Railroad Garden, 119 Red Cross Street

In 1959, Wilmington was on the verge of losing its largest employer: the Atlantic Coast Line Railroad headquarters was being moved its to Jacksonville, Florida. Festival president Walker Taylor III saw flowers as good ambassadors: "To promote good will for the city of Wilmington is the primary interest of this year's Azalea Festival. We want to let people know that Wilmington is a good place for industrial sites and an excellent place for their employees to live."

Debra Paget and Ronald Reagan posed at Orton for Hugh Morton, in 1959. (Cape Fear Museum)

Ronald Reagan, another master of the positive, was a visiting celebrity in 1959, and he became the festival's "proudest son." Though he visited Wilmington as a spokesperson for General Electric, he had already made his mark and fortune as a movie star of upper echelon Hollywood. Festival organizer Hugh Morton remembers when he, Ronald Reagan, and N. C. Governor Luther Hodges shared some private moments on the porch of Cape Fear Country Club. "Ronald Reagan asked Gov. Hodges if he, a prominent businessman, was happy in politics. Well, Governor Hodges gave him quite a speech on the value of public service. In fact, he stressed that he thought all successful businesspeople owed it to society to put in some time as public servants. Ronald

Mary Scott Bethune and Mrs. James Lamberson stand by while Debra Paget cuts the ribbon at the Rehder Garden. (The Henry B. Rehder Collection)

Reagan listened intently and I really think he changed directions that day."

Of the male celebrities, the future president took second billing that year. Azalea King John Sutton was considered the star, having been featured in "Amazon Trader," "Sangaree," and "Bride of Vengeance."

Debra Paget was queen of the Azalea Festival in 1959, the year the festival, previously local, became an official state celebration. Ms. Paget, who had a part in "Princess of the Nile" and "The Ten Commandments," also starred as Elvis Presley's wife in the 1956 hit "Love Me Tender." Debra Paget opened the garden club tour at the home of Mr. and Mrs. Henry B. Rehder, where she was duly impressed with the floral fireworks. After Mrs. A. B. Bethune, chairman of the festival tour, presented her with a basket containing a vibrant blooming azalea, the Queen said, "You know, this garden looks just like a movie set…only they never did it quite like this."

Debra Paget retired from show business a few years later and married Houston oilman Louis C. Kung, Jr.

Clarence Jones, legendary longtime gardener at Orton Plantation, holds a sparrow. Mr. Jones taught the bird to eat from his hand, a rare accomplishment caught on film by Betsy L. Sprunt. Mr. Jones's gardening accomplishments have been chronicled in several national magazines.

Tookie Bethune (on far right) leads the way down South Third Street in the 1959 parade.

Mrs. E. M. McEachern

President	Mrs. E. M. McEachern
First Vice President	Mrs. John K. Ward
Second Vice President	Mrs. W. K. Stewart, Jr.
Third Vice President	Mrs. James Lamberson
Recording Secretary	Mrs. E. L. Ward
Corresponding Secretary	Mrs. William Head
Treasurer	Mrs. F. A. Matthes, Jr.
Historian	Mrs. C. M. McKinnon
Librarian	Mrs. Charles Bakaert
Parliamentarian	Mrs. J. E. Holton, Jr.

WORK OF THE DEPARTMENTS

<u>Conservation</u> - Preservation and protection of Wild Flowers was the topic for the department. Mrs. C. M. Appleberry, Chairman.

Mrs. Appleberry planned two full trips   one a wild flower identification trip, and , second, a shell collecting field trip, which was followed by a workshop to teach the making of shell water lilies. The February meeting topic was the "Pageant of Wildflowers," sponsored by the department.

Cape Fear Garden Club yearbook courtesy of New Hanover County Public Library.

Cape Fear Garden Club contributed generously to the Kiddie Zoo at Greenfield Lake, shown here in 1958. (Cape Fear Museum)

Linda Christian, Queen Azalea XIII, was more than beautiful. The brainy actress spoke seven languages. (Joanne and Chick Mathis)

1960

Once again, April Fool's Day proved to be tricky for the garden club. Mrs. C. D. Maffitt, tour chairman, planned the tour, with the ribbon cutting set for the Burns garden. But unusually bitter weather caused the tour's cancellation. There wasn't even a decent azalea bloom for the Queen's corsage so actress Linda Christian wore a camellia instead.

However, in other ways, the 13th festival proved lucky. Linda Christian, who was discovered by Errol Flynn, once was married to Tyrone Power, and was the very first "Bond Girl" (Ian Fleming's "Casino Royale"), was a

very popular queen. Merv Griffin was the master of ceremonies. Charles Laughton, whose movie "Spartacus" came out the same year, met and became a fan of Wilmington artist Claude Howell.

And there were dances for many music tastes and age groups. Wilmington native Charlie Daniels appeared at the Ocean Plaza Ballroom at Carolina Beach. The Oleander Company hosted a "street dance" at Hanover Center, and Lumina was the site of the Teenage Princess Dance. But perhaps the most important thing about the 1960 festival was a ceremony on College Road in the midst of a pine forest: officials of Wilmington College broke ground for a new campus.

The decade of the sixties would prove to be a good one for local gardening projects. The twenty area garden clubs that made up the Cape Fear Garden Council were responsible for approximately a thousand plantings at various local sites. Cape Fear Garden Club, being the largest group, dominated many of the endeavors.

Garden Club member Sadie Stadiem Block paused one rare moment, about 1960, to smell the roses. A diligent gardner, she often began working in her garden at 711 Forest Hills Drive at dawn and worked until late afterrnoon. (Lower Cape Fear Historical Society)

This Orton Plantation float, about 1960, featured Wilmington girls Libby Sprunt (left), Beth Tillery, Susan Bekaert (front left) and Kristen Rehder (front right). (*Star News* photograph, courtesy of Betsy and Kenneth Sprunt)

1961

Mrs. Miriam Burns, 1417 Hawthorne Road

Mr. and Mrs. J. E. Furr, 702 Northern Boulevard

Dr. and Mrs. Graham Barefoot, 206 Forest Hills Drive

Miss Allie Morris Fechtig, 123 West Renovah Circle

Mrs. P. R. Smith, 615 Forest Hills Drive

Mr. and Mrs. Ralph Bertram Williams, 708 Forest Hills Drive

Mr. and Mrs. Peter Browne Ruffin, 753 Forest Hills Drive

Col and Mrs. George W. Gillette, 1236 South Live Oak Parkway

Mr. Martin Pearsall, Airlie Road (Gray Gables)

Mr. and Mrs. Henry B. Rehder, 2217 Oleander Drive

The 1961 festival was scheduled to begin a little later in April. Everyone, especially chairman Mrs. W. F. Elliott, was thrilled when all the flowers seemed to appear just in time for the garden club tour. Throngs of onlookers watched as the queen cut the ribbon at Miriam Burns's residence on Hawthorne Road. Records numbers of guests toured the gardens, many having arrived on buses that left from the Community Center Flower Show, at Second and Orange streets. Wilmington was the first stop on the Horticultural Society of New York's 1961 East Coast Tour, and many guests marveled at a prize-winning camellia entered by Mrs. Kenneth Sprunt.

Mrs. E. M. McEachern, an esteemed local historian, was president of the garden club in 1961. It is interesting that during her administration, there is a prelude to what would become the Azalea Belle tradition. According to the Cape Fear Garden Council Scrapbook for that year, "Twelve young women dressed in antebellum costumes waited beyond the ribbon lining the entrance to greet the queen."

Shelly Fabares, who had gained fame as Dr. Stone's daughter on "The Donna Reed Show," was Queen XIV. Nick Adams, who starred as Johnny Yuma on a primetime show called "The Rebel," got most of the attention from the young teenage girls. But much of the female audience had eyes only for John Larkin, who played Mike Kerr on "The Edge of Night." According to Joe Bennett, who worked at the airport, there had never been anything like the reaction he elicited. "My brother, Jack, was flying him out and there were so many women around the plane, they had to wait. One of the fellows that worked out there was a perfect mimic and could talk just

Actor John Larkin chats with J. E. L. Wade, one of Wilmington's strongest promoters. (Cape Fear Museum)

like John Larkin. We called one of those women to the phone and told her someone wanted to speak to her. As soon as she heard the voice, she fainted."

Azalea blooms at Airlie, about 1948. (Photo by Gilliam K. Horton, Josephine Corbett Horton Collection)

Mr. and Mrs. Robert F. Cameron, 726 Forest Hills Drive

Mr. and Mrs. Daniel H. Penton, 1007 Live Oak Parkway

Mrs. William G. Broadfoot, 148 Colonial Drive

Mr. and Mrs. Ralph Bertram Williams, 708 Forest Hills Drive

Mr. and Mrs. A. G. Millican, 2304 Metts Avenue

Mrs. Jessie Kenan Wise, 1709 Market Street

Miss Allie Morris Fechtig, 132 West Renovah Circle

Mrs. P. R. Smith, 615 Forest Hills Drive

Mrs. Edwin A. Harriss, Tremont at Eshcol, Masonboro Sound

Festival president Allan Strange had a very good year. The ribbon cutting at Robert and Libby Cameron's garden was beautiful. And it was at Bruce and Dan Cameron's annual Azalea Festival party at the Cape Fear Club in 1962 that Richmond resident Lawrence Lewis announced that he was going to build a "million-dollar hotel" at Wrightsville Beach: the Blockade Runner.

Azalea Queen Whitney Blake had a most recognizable face: She played Dorothy Baxter on the prime-time show "Hazel," a role that would last from 1961 until 1965. Later she made guest appearances on many other television shows, including "Batman," in 1967, where she played the role of Amber Forever. Whitney Blake also created the hit series "One Day at a Time," popular in the 1970s and '80s.

Whitney Blake was crowned at midnight, at Lumina, but sadly the old pavilion was on its way out as an Azalea Festival venue. "Barny Lumina" is what a *News and Observer* writer called it. "The dance floor of rough boards is a hazard to high-heel wearing women. But many, knowing the fact, wore ballet slippers." The preservation movement would come to Wrightsville Beach too late to save the ragged treasure. Lumina was razed in 1973.

Cape Fear Garden Club invested money and time on the grounds of the USS *North Carolina* and a garden that once was part of St. John's Lodge, at 110 Orange Street. During the 1962 Azalea Festival, St. John's Art Gallery opened for the first time in the old Masonic Lodge building. In its first year, shows featured the work of 97 artists. Today the gallery has relocated and has a new name: The Louise Wells Cameron Art Museum.

Other attractions in 1962 included a regatta at Wrightsville Beach and the Azalea Festival Masquerade Ball, held for the teenagers. The Azalea Festival Horse Show at Hugh MacRae Park attracted 1000 spectators. Some of the winners that year included Wylie Smith, Libby Sprunt, Cae Emerson, Jane Fine, Rachel MacRae, Allison Gregorie, Hardy Parker, Jr., and Susan Bekaert.

Queen Whitney Blake received mountains of flowers during her brief stay in Wilmington. Before leaving her hotel room for the airport, she was careful to arrange for send-ing them to the children's wards at James Walker Memorial Hospital.

1963

Mr. F. D. Edwards, 1229 South Live Oak Parkway

Mr. and Mrs. Robert F. Cameron, 726 Forest Hills Drive

Mr. and Mrs. Peter Browne Ruffin, 753 Forest Hills Drive

Mrs. Miriam Burns, 1417 Hawthorne Road

Mr. and Mrs. Henry B. Rehder, 2217 Oleander Drive

Queen Nancy Malone, who had a starring role in "The Naked City," did a bright job of reigning over a showery festival with an abbreviated list of gardens. Nancy Malone made appearances on other primetime shows, as well as keeping a regular role on "The Guiding Light" in the 1960s. But, like many azalea queens, Nancy Malone had a lot of intelligence to go along with her beauty and acting ability. In more recent years she has been a director of a long string of major television shows, including "Touched by an Angel," "Melrose Place," and "Dawson's Creek," a Columbia TriStar production that was filmed mostly in Wilmington for six seasons, ending in 2003.

James Drury starred in "The Virginian," a television show that lasted from 1962 until 1971. Joe E. Ross (Gunther Toody) and Fred Gwynne (Francis Muldoon), stars of the popular show "Car 54, Where are You?" were slated

James Drury, Annie Gray Sprunt, Joe E. Ross, and Beverly Stark pose at Orton Plantation. (Orton Collection)

guests. Francis Muldoon became better known in 1964 as Fred in "The Munsters."

As part of the 1963 festival season, Cape Fear Garden Club participated in a Garden Council program honoring the Greenfield Drive Association. That year, six Eleyi crab apple trees were planted at the Greenfield Amphitheater. Many garden club members also attended the Azalea Festival worship service at St. Philip's Church in Brunswick Town.

Also in 1963, Cape Fear Garden Club dedicated the Carl Rehder Memorial Garden, an arboretum and native plant garden established on a four-acre tract adjacent to Lake Forest School. Mr. Rehder worked with the club on several school garden projects but earned local fame for his efforts to teach needy families to grow their own produce. His work during the Great Depression earned him state recognition.

1964

Col. and Mrs. George W. Gillette, 1236 South Live Oak Parkway

The E. F. Beale, R. P. Huffman, and G. W. Ross families: A Neighborhood Garden on Metts Avenue

Mr. and Mrs. J. Kyle Bannerman, 2512 Market Street

Mr. and Mrs. Ralph Bertram Williams, 708 Forest Hills Drive

Mr. and Mrs. Peter Browne Ruffin, 753 Forest Hills Drive

Mr. F. D. Edwards, 1229 South Live Oak Parkway

Oakdale Cemetery, 520 North 15th Street

Greenfield Lake, South Third Street

Queen Abby Dalton, who cut the ribbon at the Gillette garden, was known for her regular appearances on "The Joey Bishop Show" and "The Jonathan Winters Show." She continued her work for many years, appearing in "Falcon Crest" and "Murder, She Wrote."

Mrs. Colin F. Churchill, chairman of the tour, announced that the neighborhood garden near Forest Hills Drive on Metts Avenue would be lighted and open for evening tours. The garden club also hosted a coffee hour at

This is to certify that

MRS. J. HENRY GERDES ....is an

AMATEUR ACCREDITED JUDGE OF FLOWER SHOWS
and has also passed the required Refresher Course.

NATIONAL COUNCIL OF STATE GARDEN CLUBS, INC.

Violet Gose

Date 5-24-1964 .... President

Cape Fear Country Club where virtually every woman wore a hat decorated with flowers.

Much attention was paid to Michael

Garden Lovers enjoying an event at 711 Forest Hills Drive, about 1964.

Landon and Frankie Avalon. Landon, who starred early in his career in the movie "I Was a Teenage Werewolf," had phenomenal success with the smaller screen. In 1964, he played Little Joe in the long-running show "Bonanza." After "Bonanza" was cancelled, he starred in "Little House on the Prairie" and "Highway to Heaven," among other shows. Frankie Avalon, the ebullient star of many beach-blanket movies, was also a singer. His rendition of "Venus" still stands as the song's standard and takes on a dual meaning in view of Wilmington's signature plant.

By the mid 1960s, festival garden tours were making more money than the founders probably envisioned. Their success enabled Cape Fear Garden Club to contribute to many projects of beautification and natural preservation. Cape Fear Garden Club also contributed

garden books to New Hanover County Public Library, an ongoing practice that, according to librarian Beverly Tetterton, "has made the local gardening collection a treasure."

A partial list of garden club and garden council projects from the 1960s includes plantings at the YWCA on Market Street (The Gilchrist House); dedication of a four-acre garden at Greenfield Lake (adjacent to Lake Forest School) in honor of home-garden enthusiast extraordinaire Carl Rehder; beautification of the Bijou theater site; the creation of a Rose Garden at Greenfield Lake; dogwood trees planted as a tribute to Mayor O.O. Allsbrook, near 16th and Church streets; annual Arbor Day ceremonies; planting 300 shrubs and trees near the Fragrance Garden at Greenfield Lake; placing two plaques at Greenfield Lake, dedicated May 8, 1964, one bearing the words of Edna St. Vincent Millay's "Renascence"; and planting five Kwanzan Cherry trees in honor of J.E.L. Wade, the local public servant who did the most to help Greenfield Lake.

Also, several members of the garden

Greenfield Lake. (Photo by Freda Wilkins)

club were pursuing specialized garden activities. Effie Barefoot Burney and Marie Rehder Gerdes were quietly continuing their study of the ancient oriental flower art of Ikebana. Mrs. Burney and Mrs. Gerdes were both invited to enter arrangements in the 1965 Charlotte Merchandise Mart show. Mrs. Burney entered "Chinese Inspiration," featuring a distinctive container, a temple dog figurine, and a teakwood stand from Hong Kong. Mrs. Gerdes entered an arrangement to accompany "A Memory," her 1925 oil painting of gardenias by Elisabeth Chant for an exhibition at the YWCA, where the first Cape Fear Garden Club meetings were held.

The 1964 celebration gave a nod to a local festival and parade precursor, the Feast of Pirates, held from 1927 until 1929. However, the pirate extravaganza featured fewer flowers and more unchecked consumption of party beverages. General rowdiness and some 1920s-style exotic dancing on Third Street combined with the onset of the Great Depression to squelch the Feast of Pirates. Wilmington's wry critic, artist Claude Howell, witnessed both festivals and described the Azalea Festival as rather "saccharine," blaming it on the town's reaction to the earlier countywide party.

But, among others, jazz lovers would disagree with Mr. Howell. Billy Rehder and Paul T. Marshburn were always busy adding musical sensuality to the festival. In time, they recruited such jazz greats as Cab Calloway, Billy Butterfield, and Bob Haggard for festival performances.

1965

Mr. and Mrs. Henry B. Rehder, 2217 Oleander Drive

Mr. and Mrs. Garvin D. Faulkner, 5 North 20th Street

Mrs. Miriam Burns, 1417 Hawthorne Road

Mr. and Mrs. Peter Browne Ruffin, 753 Forest Hills Drive

Mr. and Mrs. J. Kyle Bannerman, 2512 Market Street

The Beale-Huffman-Ross Neighborhood Garden, Metts Avenue

Mrs. Colin Churchill continued as chairman of the garden tour and oversaw the debut of the Battleship Sound and Light Show as a tour feature. The dramatic display told the story of the great ship to an audience seated on bleachers on Eagle's Island. The gleaming vessel bathed in bright colored lights and the shock of the simulated sound of an explosion offered some tiny diversion from the discomfort of the island's famous mosquitoes.

Hugh Morton, who spearheaded the drive to bring the Battleship North Carolina to Wilmington, remembers the very first sound and light show. "They used part of a stick of dynamite for the torpedo show. That first night, they were not sure how much to use for the desired effect. Well, they used too much and there were dead fish all over the place."

Queen Patricia Blair had starring roles in

prime-time television shows "The Rifleman" and "Daniel Boone." Ed Ames, who also played her on-screen husband Daniel Boone, was one of the festival's Renaissance men. He began his career familially as one of the singing Ames Brothers, but went solo as a musician while pursuing television, theater, and movie roles – as well as acquiring a graduate degree and maintaining an excellent game of tennis.

By 1965, 700 volunteers were involved in producing a celebration, an activity that had already honed the skills of established leaders and trained new participants to work toward the public good. A system of earning the right to be president was firmly in place; it bred discipline and patience as newcomers to the festival commission worked their way to the top.

White perfume: Gardenias at Airlie. (Photo by Gilliam K. Horton, courtesy of Josephine Corbett Horton)

1966

Mrs. Miriam Burns, 1417 Hawthorne Road

Col. and Mrs. George W. Gillette, 1236 South Live Oak Parkway

Mr. F. D. Edwards, 1229 South Live Oak Parkway

Mrs. P. R. Smith, 615 Forest Hills Drive

Mr. and Mrs. J. Kyle Bannerman, 2512 Market Street

Mr. and Mrs. Garvin D. Faulkner, 5 North 20th Street

The Beale, Huffman, Ross Neighborhood Garden, Metts Avenue

In 1966 Wilmington was feeling good about itself. Dan Cameron, a festival stalwart, played a leading part in obtaining Wilmington's status that year as an "All-American City." After years of hard work through the Committee of 100, Cameron and many other local leaders had succeeded in bringing business into Wilmington after the Atlantic Coast Line Railroad announced its departure for Jacksonville, Florida. Wilmington received national recognition for its new accomplishments. Hugh Morton and Dan Cameron were two of the 30 civic leaders who flew to St. Louis to campaign for the city's honor.

Even the *Chicago Tribune* took note of the City of a Million Azaleas. "Come April and

May, there's no place I'd rather be than smack in the middle of North Carolina's storied old plantation country at Wilmington," wrote Ken Ferguson. But it wasn't just the grand old estates that brought accolades. The twentieth-century gardens on the 1966 tour brought lavish praise from tour takers, and Mr. and Mrs. Gar Faulkner's garden won compliments for style and grace.

Cape Fear Azaleas   (Photo by Gilliam K. Horton, courtesy of Josephine Corbett Horton)

Miriam Burns's garden was the site of the ribbon cutting, an event covered live for the first time by the local NBC affiliate, WECT-TV. According to a garden club correspondent, "This was televised to a two-state audience." Mrs. Burns's son, James Moss Burns, Jr., was a long-time employee of the station.

Ulla Stromstedt, who starred in "Flipper," was Queen Azalea XIX. Hugh Morton was impressed with the queen's conversation and told a reporter, years later, that, like the first queen, Jacqueline White, Ulla Stromstedt "had sense as well as being pretty."

Larry Storch of "F Troop," singer Barry Sadler, orchestra director Mitch Miller, and country star Tex Ritter pleased the crowds. Country musician George Hamilton IV was

especially good at mingling with the crowds and received such a welcome that he returned in 1970.

As a courtesy to the public, tour chairman Mrs. W. L. Grant announced that the following weekend 31 private gardens would be open. Azaleas were at their peak and the tour was free. With the exception of Clarendon Plantation in Brunswick County and the Alexander Sprunt Garden at 1615 Chestnut Street, most were garden tour regulars.

In a prelude to the festival, Arbor Day ceremonies at Greenfield Lake featured several Cape Fear Garden Club giants. Marie Rehder Gerdes, city Arbor Day chairman, presented gifts from the Cape Fear Garden Council: eight Bradford pears for the Municipal Rose Garden. The trees were given in honor of the men who created Community Drive, the five-mile road around the lake. Several men present at the 1966 ceremony had earned desperately needed income during the Depression through the generosity of those employed citizens who gave up one day's pay each month to help Wilmington avoid the creation of breadlines.

Mrs. Gerdes, the N. C. Garden Club Collect author and general philosopher of the garden club, had a lifelong passion for trees. In addition to spearheading club tree projects, she carried on her own programs, most notably transplanting dogwood trees from her own yard to Forest Hills School. Her Arbor Day comments on trees and development constitute a talk that was ahead of its time:

"In a day when we see all around us so much tearing up and mutilation of the land, a lot of it done because we are in a hurry to get something built at once, or at any cost, it is refreshing to pause and consider how necessary it is to put something back to rejoice the hearts of those coming after us."

"Now these trees which we give to our city today are not just shade trees, as necessary as they are in the heat of summer, but these bear flowers and fruit as well, to make us mindful of our Creator's love for his children in giving us beauty to nourish our spirits as well as our bodies."

Mrs. Helen McCarl delivered the conclusion by posing a question that has no perfect answer. "It is surprising to realize that our memory has been so short as to allow the name commemorating this wonderful spirit of cooperation to be changed in recent years to one so commonplace and meaningless – Lake Shore Drive – which bears no relation to its thrilling history."

The Cloister Garden, given by one of the club's members, Mrs. Rachel Cameron MacRae, to St. James Episcopal Church, was accepted by Mrs. Roscoe McMillan, National Chairman of Memorial Gardens. (New Hanover County Public Library)

1967

The Burns-Stokes Garden, 1417 Hawthorne Road

Mr. and Mrs. Ralph Bertram Williams, 708 Forest Hills Drive

Mr. and Mrs. Robert Kallman 812 Forest Hills Drive

Dr. and Mrs. Calvin MacKay, 104 W. Renovah Circle

Mr. and Mrs. Oliver Hutaff, 2719 Mimosa Place

Mr. and Mrs. W. A. Fonvielle, Jr., 208 Wayne Drive

Mr. and Mrs. Daniel H. Penton, 1007 Live Oak Parkway

Mrs. P. R. Smith, 615 Forest Hills Drive

Miss Allie Morris Fechtig, 123 W. Renovah Circle

Mrs. William G. Broadfoot, 148 Colonial Drive

Mr. F. D. Edwards, 1229 South Live Oak Parkway

Col. and Mrs. George W. Gillette, 1236 South Live Oak Parkway

Mr. and Mrs. Henry B. Rehder, 2217 Oleander Drive

Plantation Gardens, Metts Avenue

Mrs. Sidney V. Allen chaired the tour in 1967, a lengthy pilgrimage that lasted through two weekends. Keeping up a new tradition, several gardens were open at night, this year the Plantation and the Fonvielle gardens.

The Stokes-Burns Garden had been recently enhanced by Miriam Burns Stokes's son, Jim Burns. He added walkways, plants, and graceful Italian statuary, a perfect complement to the balustrade already in place, a treasure Mrs. Stokes rescued from the ruins of Pembroke Jones's lodge at what is now Landfall.

Melody Patterson of "F Troop" was

Funds from the 1967 tour went toward improvements to the Fragrance Garden, a three-year project that was dedicated March 30, 1967. The *Star News* said the aromatic space was "the club's outstanding gift to the Blind and to the City of Wilmington." (New Hanover County Public Library)

1968

Dr. and Mrs. Raymond Grove, 1400 South Live Oak Parkway

Mr. and Mrs. Peter Browne Ruffin, 753 Forest Hills Drive

Mr. and Mrs. Robert Kallman, 812 Forest Hills Drive

Mr. and Mrs. Ralph Bertram Williams, 708 Forest Hills Drive

Mrs. P. R. Smith, 615 Forest Hills Drive

Dr. and Mrs. Calvin MacKay, 104 West Renovah Circle

Mr. and Mrs. Garvin D. Faulkner, 5 North 20th Street

Mr. and Mrs. J. E. Furr, 702 Northern Boulevard

Melody Patterson, 1967. (Henry B. Rehder Collection)

Azalea Festival Queen in 1967.

In 1967, funds from Cape Fear Garden Club were earmarked for the beautification of a new senior high school: John T. Hoggard. Additional contributions were channeled to Greenfield Lake where the club's Fragrance Garden, an Eden for the blind, was dedicated on March 30, 1967.

Members of the Cape Fear Garden Club Council picnic at Brunswick Town, September 20, 1968. (New Hanover County Public Library)

Col. and Mrs. George W. Gillette, 1236 South Live Oak Parkway

Mr. and Mrs. Daniel H. Penton, 1007 Live Oak Parkway

Plantation Gardens, Metts Avenue

Mrs. Sidney V. Allen served again as chairman of the tour. By 1968, azaleas were beginning to give way a bit to more variety. The MacKay garden on Renovah Circle featured orange trees in an enclosed courtyard. The Furr garden, on the northwest corner of Northern Boulevard and Harrison Street, in Sunset Park, featured interesting borders, unusual terrain and the gentle sounds of lambs coming from Mr. and Mrs. Furr's multi-acre sheep farm.

Linda Cristal, star of television's "High Chaparral," wowed the crowds with her Argentinean good looks. Comedian Rich Little, singer Vaughn Monroe, and Emmett Kelly, Jr., created interesting texture for the 1968 festival.

Kenneth Sprunt introduces Queen Linda Cristal to the wonders of Orton's camellias.

1969

Mr. and Mrs. Herbert Bluethenthal, 1704 Market Street

The Kenan House, 1705 Market Street

Mr. and Mrs. Charles Lowrimore, Jr., 2806 Highland Drive

Mr. and Mrs. Robert Kallman, 812 Forest Hills Drive

Mr. and Mrs. Ralph Bertram Williams, 708 Forest Hills Drive

Col. and Mrs. George W. Gillette, 1236 South Live Oak Parkway

Mr. and Mrs. Henry B. Rehder, 2217 Oleander Drive

Dr. and Mrs. Calvin MacKay, 104 West Renovah Circle

Plantation Gardens, Metts Avenue

Kenan House, once the residence of Cape Fear Garden Club member Sarah Graham Kenan, has been home to University of North Carolina at Wilmington chancellors since 1969. (Cape Fear Museum)

The year 1969 gave the festival a new and unexpected piece of news. While on a flight to Miami to recruit golfers for the Azalea Open, Chris E. "Gene" Fonvielle and James B. "Bunny" Hines were hijacked to Cuba. After a brief interrogation and a snack, the passengers were sent back to the U. S., unharmed and, for the most part, unshaken.

Teenage girls in Colonial dress had occasionally dressed up some of the gardens during the Pilgrimage and the early Cape Fear Garden Club tours. Historian Leora Hiatt McEachern, who was club president from 1959 until 1961, encouraged the practice. But 1969 was the first year that Azalea Belles appeared as a coordinated group. Mrs. Harley Vance, president of the club, is credited for making belledom an official part of the Azalea Festival.

Another belle enthusiast was Mrs. W. A. Fonvielle, who lived on Wayne Drive. The first year they presented seven young ladies: Marsha Blake, Jean Burdette, Wanda Johnson, Beth Chadwick, Ginger King, Kathie White, and Pamela Wood.

Five of the original belles were daughters of members of the club, including Marsha Blake, whose mother, Louise, was chairman of the 1969 tour. The hoops most of the girls wore belonged to their mothers, left over from the Cape Fear Confederate Ball held at Cape Fear Country Club, April 15, 1962.

Cape Fear Garden Club membership continues to be a qualifying factor for daughters and granddaughters in becoming an Azalea Belle. And years of work equity can boost a

Chris Noel waits for a *Life* magazine photo shoot tobegin at Greenfield Lake. Both Milton Green, renowned for his photos of Marilyn Monroe, and Hugh Morton, North Carolina's most famous photographer, supplied pictures of the Azalea Festival to *Life*. (Photo by Hugh Morton)

member's descendants from ordinary belle to local magazine cover status. Today however, any girl who is at least 16 years old, in the 10th, 11th, or 12th grade, and is endorsed by a member of Cape Fear Garden Club, is a potential wearer-of-the-hoop. The post is open to girls of all creeds.

Wandering through some of the gardens that year was David Hartman, an actor who was particularly good at mingling with the crowds. Hartman, a Duke graduate, appeared in "The Virginian" and "Lucas Tanner" before beginning a 12-year stint as host of "Good Morning America," in 1972. Also here for the 1969 festival were Alejandro Rey, Henny Youngman,

Al Hirt, and Queen Chris Noel, a beach movie actress who also made entertainment tours in Viet Nam.

Guests were especially eager to tour the Kenan House garden, a formal space that creates perpetual decoration through a glass wall at the rear of the mansion. Mrs. Kenan acquired marble benches, a fountain and numerous other decorations for the garden on a trip to Italy. A large privacy wall creates a handsome boundary for the property.

The 1969 tour brought rich proceeds that Cape Fear Garden Club funnelled into projects to beautify the grounds of local housing projects, as well as Empie Park and Cape Fear Technical Institute.

The first official Azalea Belles, 1969.

(Star News photographs, Jo Chadwick)

1970

The Stokes-Burns Garden, 1417 Hawthorne Road

Dr. and Mrs. John Codington, 624 Forest Hills Drive

Mr. Claud Efird, 807 Forest Hills Drive

Mr. and Mrs. George Caplan, 321 Renovah Circle

Mr. and Mrs. J. D. Causey, 2037 Shirley Road

Mr. and Mrs. Oliver Hutaff, 2719 Mimosa Place

Mrs. P. R. Smith, 615 Forest Hills Drive

Mr. and Mrs. J. Henry Gerdes, 722 Forest Hills Drive

Mr. and Mrs. Peter Browne Ruffin, 753 Forest Hills Drive

Plantation Gardens, Metts Avenue

Mrs. G. Watts Farthing chaired the 1970 tour. Once again, the Queen's garden party and ribbon cutting took place at Stokes-Burns. Queen Karen Jensen starred in the 1960s spy spoof "Out of Sight" and played a part in "The Wild Wild West," a long-running television show. Though everyone else seemed impressed with the pretty blonde queen, the queen herself was impressed with Mrs. Lizzie Jones, a black woman who had been the Sprunt family cook for 50 years. The two women talked recipes and Mrs. Jones shared some of her kitchen secrets.

Orton employee Lizzie Jones with Karen Jensen, 1970. (*Star News* photograph, Betsy and Kenneth Sprunt)

Azalea Belles proliferated in 1970 with 25 teenage girls flanking the ribbon cutting and then dividing their hours among the ten gardens.

Visiting writer Marion Gregory attended the festival party at Orton Plantation that year and offered a little different interpretation of Wilmington's busiest weekend in *The News and Observer.* "For the social set of the city, heavily Episcopalian, it's a chance to burst out of the barren Lenten season with a full weekend of parties."

Mrs. Ruffin wrote in her diary: "There were streams of people in my garden each day during the tour and many continued to come the next week. The compliments were profuse and made all my work worthwhile." (Peter Browne Ruffin Collection)

1971

Mr. and Mrs. Duval Lamdin, 711 Forest Hills Drive

Purnell-Empie House, 319 South Front Street

Mr. and Mrs. Malcolm B. Lowe, Jr., 726 Forest Hills Drive

Mr. and Mrs. Charles Lowrimore, Jr., 2806 Highland Drive

Mr. and Mrs. J. D. Causey, 2037 Shirley Road

Kenan House, 1705 Market Street

Gray Gables, Airlie Road

Plantation Gardens, Metts Avenue

The tour, under the leadership of chairman Mrs. G. Watts Farthing, brought about some garden tour changes that marked time and new ownership. The Cameron home at 726 Forest Hills Drive had been sold to someone outside the family: Mr. and Mrs. Malcolm B. Lowe. Also, the garden at 711 Forest Hills Drive was in new hands. Established in 1940 by Mrs. Nathan E. Block, she also created a large lathe house along the railroad track that bordered their property. Working daily, wearing heels and jewelry, she created a beautiful array of camellias to accompany the azaleas that blanketed her land. Though a member of the garden club, Sadie Block was a perfectionist who never considered her garden "finished." She declined numerous invitations to

be on the garden tour. However, she won garden club applause in 1956 when she created the annual beautification project by requesting that developer Hugh MacRae II dedicate a triangle of land at the intersection of Wrightsville Avenue and Forest Hills Drive as green space. Mr. MacRae agreed and the city of Wilmington granted permission for water connections. Cape Fear Garden Club planted azaleas and added benches to the property. The project won an award from the Garden Club of North Carolina, Inc.

In 1969, Sadie and Nathan Block sold their home to Mr. and Mrs. Duval Lamdin. The Lamdin family, who hosted the ribbon cutting and garden party in 1971, maintained the garden until 1990, when the house was sold to Dr. and Mrs. William Eason. Since that time, Marie Eason has improved the established garden, adding many new species and creative designs.

Hannah Block, Sadie Block's sister-in-law and a perennial festival supporter, created a new situation altogether when she left her Forest Hills house and moved downtown. Though revitalization of the historic district had begun, few people were willing to make it their primary residence. Hannah Block opened her garden to the public in 1971, displaying an attractively manicured space that had only recently been a mass of weeds and litter.

In 1971, Cape Fear Garden Club lost a generous member, Miss Allie M. Fechtig (1896-1971). She served two terms as president and was the first president of the Cape Fear Garden Council. An incorporator of Greenfield Lake, she gave generously through the years to beautification projects there. In 1971, the council planted a tree at Greenfield in her memory.

Phyllis Davis, who enjoyed a five-season run on "Love American Style" was Queen Azalea XXIII. Like so many other early azalea queens, Ms. Davis had been in Elvis movies before landing a regular television contract. After "Love American Style," she took the role of Beatrice in the Aaron Spelling show "Vega$" from 1978 until 1981.

1972

Mrs. P. R. Smith, 615 Forest Hills Drive

Mr. Oscar Durant, 704 North Lumina, Wrightsville Beach, a Japanese garden

Mr. and Mrs. Robert Kallman, 812 Forest Hills Drive

Mr. and Mrs. J. D. Causey, 2037 Shirley Road

Mr. and Mrs. Henry B. Rehder, 2217 Oleander Drive

The Purnell-Empie House, 319 South Front Street

Gray Gables, Airlie Road

Plantation Gardens, Metts Avenue
Chairman Mrs. Harley Vance proudly announced the tour's first beach garden and organized the ribbon cutting at the P. R. Smith residence.
Jacqueline White, the first Azalea Festival queen, returned for the 1972 event and crowned Ann Elder Queen Azalea XXV. Ms. Elder, an actress, was a regular on "The Smothers Brothers Show" and had a current role on "Laugh-In." Possessed of a thorn-sharp

wit, she kept things lively, then went on to win several Emmy Awards in the comedy category for writing. Demond Wilson of "Sanford and Son," Pat Boone, The New Christy Minstrels, and Ann Davis of the "Brady Bunch" were also guests.
The 1972 tour netted $1,500. The funds

Alison Block was a belle before hoops were considered a strict requirement. Her grandmother, Sadie Block, had so many garden club ties that belledom was Alison's festival destiny.

were distributed to the garden at St. James Church, the YWCA Building Fund, and the Girls' Club. Funds from the previous year were used to plant a garden at St. John's Art Gallery at 110 Orange Street. Tulips, ivy, and boxwood were planted around the old well and abandoned flower beds were restored.

1973

Dr. and Mrs. Edwin Wells, 2205 Gillette Drive

Mr. and Mrs. Richard P. Reagan, 2117 Gillette Drive

Mr. and Mrs. Richard A Ryder, 2029 Bradley Drive

Mr. and Mrs. Harry Burke, Figure Eight Island

Mr. and Mrs. Horace T. King, Figure Eight Island

Mr. and Mrs. F. M. Southerland, Figure Eight Island

Mr. Oscar Durant, 704 North Lumina, Wrightsville Beach, a Japanese garden

Plantation Gardens, Metts Avenue

Again Mrs. Harley Vance chaired the Cape Fear Garden Club tour. The addition of three gardens on Figure Eight Island helped make it a particularly successful year. Development was still a new thing on the pristine barrier island and access was limited usually to homeowners and their guests. Large numbers of garden visitors allowed the Garden Club to later write a substantial check to the Kiwanis Club for their "Save Greenfield Lake" fund.

Joan Van Ark, an actress trained at Yale Drama School, reigned over the 1973 Azalea Festival. She had appeared on Broadway several times, and, in 1973, she was nominated for a Tony award for her role in "School for Wives." Joan costarred in the CBS-TV series "We've Got Each Other" and the ABC-TV series "Temperatures Rising," and has appeared in numerous television guest starring roles through the years. She is perhaps best known for her role as Valerie Ewing in the CBS-TV series "Knots Landing." Reminiscing in 2003, she said that her most vivid memory of the Azalea Festival was "the police escort into town from the airport, lights flashing, sirens blaring! It was a 'touch' that I wanted to go on forever!"

A visit by the legendary bandleader Cab Calloway was a special gift to the festival that year. Mr. Calloway, who once made nightly broadcasts from New York's Cotton Club, was world-renowned veteran performer when he made his Azalea Festival appearance.

Elise Wilkins (Wilson) and Paige Scott (Mullin) pause on the bridge at Plantation Gardens. (Photo by Freda Wilkins)

1974

Mr. and Mrs. Hugh MacRae II, 807 Forest Hills Drive

Mr. and Mrs. John K. Davis, 40 Beauregard Drive

Mr. and Mrs. Thomas McCall, 106 West Cascade Road

Keene-Brink Gardens, 5122 Clear Run Drive

Mr. and Mrs. J. Henry Gerdes, 722 Forest Hills Drive

Mr. F. D. Edwards, 1229 South Live Oak Parkway

Mr. and Mrs. Richard A. Ryder, 2029 Bradley Drive

Plantation Gardens, Metts Avenue

Mrs. Harley Vance chaired the tour that began in the garden of Bambi and Hugh MacRae. Hugh MacRae is the grandson of Mrs. Hugh MacRae, a founder of Cape Fear Garden Club. Bambi MacRae is a long-time member.

Queen Sharon Gless, a star on the television show "Marcus Welby, M.D.," cut the ribbon. Later, Ms. Gless played the role of Chris Cagney in "Cagney and Lacey" for six years and won two Emmys for her portrayal.

During the festival, the garden club also promoted a separate tour of Bald Head Island.

Sharon Gless, 1974. (Henry B. Rehder Collection)

1975

Mr. and Mrs. Horace W. Corbett, 1925 South Live Oak Parkway

Mr. and Mrs. W. K. Stewart, Jr., 2203 Marlwood Drive

Jim Burns Waterside Memorial Garden, 530 Waynick Boulevard

Burns-Stokes Garden, 1417 Hawthorne Road

Mr. and Mrs. Henry B. Rehder, 2217 Oleander Drive

Mr. and Mrs. Hugh MacRae II, 807 Forest Hills Drive

Mr. and Mrs. J. Henry Gerdes, 722 Forest Hills Drive

The Keene-Brink Gardens, 5122 Clear Run Drive

Oscar Durant, 704 North Lumina, Wrightsville Beach, a Japanese garden

Plantation Gardens, Metts Avenue

Mrs. Harley Vance again chaired the tour which began with the Horace Corbett garden. Even though the dates of the festival had been pushed forward since some of the first chilly years, nature still pulled some surprises. The Corbetts invited guests to escape the cold rain by coming inside their house and basking in the heat aound their indoor pool.

The waterside garden created by local television celebrity Jim Burns made an interesting addition to the 1975 tour. Host of a popular midday talk show, he fashioned the garden between his house on Waynick Boulevard and Banks Channel. It featured statuary, a fountain, and a memorial to 52 North Carolina families who contributed to the state's beautification.

Bob Hardesty of WECT served as Master of Ceremonies for the ribbon cutting. Azalea Queen Stephanie Braxton starred in the daytime drama "All My Children" at the time of her reign but later became a television writer.

The Rehder garden had become a perennial favorite, on the tour officially or not, because the owners posted a "Welcome" sign throughout the blooming season. The Rehders' gracious Southern house, built by Eleanor Wright Beane, came with an extra lot. Though beautifully landscaped today, it wasn't always so.

"There was a caddy path running through it, from Delgado on Wrightsville Avenue, to Cape Fear Country Club. It was a heavily wooded area," said Mr. Rehder, in 2003, "and the first time my wife, Barbara, and I worked out there, we worked for a whole day and cleared about one square yard. So we had to get a land surveyor and we got Henry Von Oesen. He came in and helped plan it. We did it little by little. That's why it's in rooms."

The Rehder garden now contains 125 camellias, as well as ornamental fruit trees and kurume, formosa, and indica azaleas. Mr. Rehder has been host to many famous people through the years, including actresses Mary Martin and Helen Hayes. National magazines have carried stories on Henry Rehder and his garden has frequently been a part of the North Carolina Garden Tour.

The majority of the proceeds from the 1975 Cape Fear Garden Club tour were given to the Bellamy Mansion Garden Fund. Cape Fear Garden Club also provided additional funding to Brunswick Town through a joint project of the Garden Club of North Carolina, Inc. In 1975, Cape Fear Garden Club was commended for its gifts to the Colonial site, assistance in establishing a nature trail, and the contribution of redwood for benches and tables.

1976

Mrs. Jack C. Thompson, 1938 South Live Oak Parkway

Mrs. William L. Walker, 1840 South Live Oak Parkway

Dr. and Mrs. Charles L. Nance, 1922 Brookhaven Road

Mr. and Mrs. William Echols, 738 Forest Hills Drive

Mrs. P. R. Smith, 615 Forest Hills Drive

Mr. and Mrs. J. Henry Gerdes, 722 Forest Hills Drive

Mr. and Mrs. Duval Lamdin, 711 Forest Hills Drive

Mr. and Mrs. John K. Davis, 40 Beauregard Drive

Kenan House, 1701 Market Street

Plantation Gardens, Metts Avenue

Marie Rehder Gerdes in her garden with sons Jon (on left) and Phillip, about 1942. Louise Harriss wrote of her: "She knows by name and ancestry of everything that grows there. ....terraced woodland with a brook traversed by rustic bridges...where wildlings live compatibly with fine varieties of camellias and azaleas."

The P. R. Smith garden, at 615 Forest Hills Drive. (Photo by Hugh Morton)

Mrs. J. J. Pence, chairman of the tour, was thankful that Mrs. Thompson's dozens of Yoshino cherry trees' narrow window of blooming time coincided with the ribbon cutting and Queen's Garden Party. Rita McLaughlin, an actress on the daytime drama "As the World Turns," was Queen Azalea XXVIII. Ted Lange, who played Isaac on "Love Boat," was a festival guest. Patti and Richard Roberts performed with their group, the World Action Singers.

1977

Mr. and Mrs. Albert F. Perry, 2511 Canterbury Road

Mr. Samuel H. Hughes, 309 Stradleigh Road

Mr. and Mrs. Richard P. Reagan, 2117 Gillette Drive

Mr. and Mrs. William Echols, 738 Forest Hills Drive

Dr. and Mrs. John Codington, 624 Forest Hills Drive

A fountain in the Burns-Stokes Garden. (Photo by Freda Wilkins)

Dr. and Mrs. Calvin MacKay, 104 West Renovah Circle

The Burns-Stokes Garden, 1417 Hawthorne Road

Mr. and Mrs. Richard A. Ryder, 2029 Bradley Drive

Plantation Gardens, Metts Avenue

Mrs. J. J. Pence chaired the 1977 tour. The crowd was unusually large at the Perry garden ribbon cutting, presented live on TV. Excessive heat had already wilted the blooms but locals and clubs' members from all over the state enjoyed seeing garden designs nonetheless.

Queen Francesca James acted on "All My Children." Celebrities from "The Lawrence Welk Show" included Thomas Netherton and Norma Zimmer, the Champagne Lady. Charlotte Stewart of "Little House on the Prairie" and comedian John Byner also made appearances.

1978

Dr. and Mrs. John F. White, 748 Forest Hills Drive

Dr. and Mrs. R. V. Fulk, 764 Forest Hills Drive

Mr. and Mrs. S. L. Marbury, 741 Forest Hills Drive

Mr. and Mrs. J. D. Causey, 2037 Shirley Road

Mr. and Mrs. Henry B. Rehder, 2217 Oleander Drive

Mr. and Mrs. Thomas McCall, 106 West Cascade Road

Mr. and Mrs. Richard A. Ryder, 2029 Bradley Drive

Plantation Gardens, Metts Avenue

Dr. and Mrs. Clayton B. Smith hosted a tea for the 32 Azalea Belles at their home on Greenville Sound. Mrs. Harley Vance continued as the belles' advisor and Mrs. J. J. Pence served again as tour chairman. Queen Nancy Addison of "Ryan's Hope" cut the ribbon at theWhites' garden. Mr. and Mrs. Sammy Davis and singer B. J. Thomas were special guests.

In 1978, Cape Fear Garden Club made a generous donation to Oakdale Cemetery for the pruning and feeding of its many dogwoods.

1979

Photographer Freda Wilkins caught this yellow swallowtail having its own azalea tour.

Dr. and Mrs. James H. Robinson, 1903 Brookhaven Road

Mr. and Mrs. Julian H. McKeithan, 1719 Fairway Drive

Mrs. S. O. Guyton, 107 North 26th Street

Mr. and Mrs. William S. Howell, 222 Masonboro Loop Road

Airlie Gardens, W. Albert Corbett family, Airlie Road

Mr. and Mrs. A. W. Blount, 2826 Columbia Avenue

The Stokes-Burns Garden, 1417 Hawthorne Road

Mrs. P. R. Smith, 615 Forest Hills Drive

Burns Waterside Garden, 530 Waynick Boulevard

Plantation Gardens, Metts Avenue

The inclusion of Airlie as an official garden club tour site delighted garden visitors.

Among the many garden tourists through Airlie that year were numerous Coast Guardsmen. Connie Parker, chairman of the 1979 tour, granted free admission to the crew of the cutter *Northwind*, and their families, as a "welcoming gesture on behalf of the entire community."

The year 1979 was the first time the Azalea Festival received statewide television promotion. Statistics collected showed that garden visitors came from 83 cities in North Carolina, as well as 30 other states and four foreign countries. Laurie Walters, who played Joanie in the nighttime television series "Eight is Enough," was queen. Debby Boone, singer of "You Light up My Life," and Pure Prairie League entertained the festival audience.

Azalea Belle champion Fanchen Vance is in the midst of her charges at the home of Toppy and Jim Robinson, in 1979. (Photo by Freda Wilkins)

The five-acre P. R. Smith Garden at 615 Forest Hills Drive proved to be a continuing favorite with festival guests. (Photo by Freda Wilkins)

1980

Dr. and Mrs. John Codington, 624 Forest Hills Drive

Mr. George W. Jones, Jr., 510 Surry Street

Mr. Henry J. MacMillan, 118 South Fourth Street

Mr. and Mrs. Stuart Y. Benson , 311 Cottage Lane

Mr. and Mrs. Eugene W. Merritt, 1209 Essex Drive

Mr.and Mrs. Donald B. Stegall, 339 Camp Wright Road

Rice fields and a short rail line from the Orton dock gave way long ago to ornamental gardens that draw tens of thousands of visitors to the old Brunswick County plantation. (Photo by Freda Wilkins)

Mr. and Mrs. William R. Smaltz, 541 South Lumina Avenue

Mrs. William L. Walker, 1840 South Live Oak Parkway

Mr. and Mrs. Albert F. Perry, 2511 Canterbury Road

Mr. and Mrs. Malcolm B. Lowe, 726 Forest Hills Drive

Roger Moore built Orton in 1725. It's still Cape Fear's grandest property. (Photo by Freda Wilkins)

Mrs. P. R. Smith, 615 Forest Hills Drive

Mr. and Mrs. Richard A. Ryder, 2029 Bradley Drive

Plantation Gardens, Metts Avenue

Tour chairman Mrs. William P. Parker, Jr., presided over the ribbon cutting at the home of Dr. and Mrs. John Codington, where azaleas and camellias grace the spacious open lawn on Forest Hills Drive. Local artist Henry J. MacMillan, who provided drawings for many early Cape Fear Garden Club flower shows, agreed to be on the tour. An especially varied group of gardens brought great interest and a ticket take of $6,000.

The 1980 Azalea Queen was Lacy Neuhaus of "From Here to Eternity." Dionne Warwick and Pete Fountain provided musical entertainment to appreciative audiences.

1981

UNC-W Pedestrian Mall

Mr. and Mrs. Ralph L. Godwin 1902 Princess Street

Mr. and Mrs. Harold W. Wells, 1217 Country Club Road

Mr. and Mrs. J. D. Causey, 2037 Shirley Road

Mr. and Mrs. Henry B. Rehder, 2217 Oleander Drive

The Stokes-Burns Garden, 1417 Hawthorne Road

Mr. and Mrs. Julian H. McKeithan, 1719 Fairway Drive

Dr. and Mrs. Raymond Grove, 1400 South Live Oak Parkway

Mr. and Mrs. William S. Howell, 222 Masonboro Loop Road

James Moss Burns, Waterside Memorial Garden, 530 Waynick Boulevard

Plantation Gardens, Metts Avenue

Chairman Eleanor Hunt organized a successful tour that made $10,000. The ribbon cutting and garden party at the UNC-W Pedestrian Mall gave a new bond to Cape Fear Garden Club and the university. Stretching back to the days when UNC-W was Wilmington College and operated in the Isaac Bear building on Market Street, members of the club shared their knowledge of nature and their monetary donations with the University of North Carolina at Wilmington.

When Wilmington College and the Wilmington Merchants Association decided to publish *A Bird's Eye View of the Tourist Training Program*, the school asked Mrs. Cecil Appleberry, Allie Morris Fechtig, and Marie Rehder Gerdes to write chapters for the small book. All three gave talks at the young college. Then, over the years, Cape Fear Garden Club did numerous beautification projects after the school moved to the College Road campus. The Pedestrian Mall, created in 1981 with two years of tour proceeds, is the club's most ambitious college project to date.

Other projects of Cape Fear Garden Club and the Garden Council for 1981 included conducting garden therapy activities at four public housing projects, beautifying seven medians on 17th Street, and planting 73 crepe myrtles and 700 shore junipers in New Hanover County.

The 1981 queen was Maureen Teefy, who played Doris in the 1980 movie "Fame." Bill and Susan Seaforth Hayes from the daytime drama "Days of our Lives" were crowd favorites. Gayle Ward, a tireless festival promoter, struck up a warm friendship with Susan Hayes that has lasted 23 years.

But it was a 77-year-old man who stole the weekend show. Bob Hope gave two performances in Trask Coliseum, both in one night. Working on a flowery stage in-the-round, Mr. Hope delivered one-liners without using cue cards and delivered different jokes for each show.

Though he needed a little assistance to navigate the small staircase leading to the stage, nothing else but the occasional fleeting outline of a back brace gave any indication that he was a septuagenarian. His performance was based on mental calisthenics worthy of Olympic status. Wilmingtonians left knowing they had actually seen a great and long-standing American patriot, a master of humor, and an international institution.

1982

James Moss Burns, Jr., (on left, in yellow pants) was a colorful television personality, a passionate gardener, and a great friend to Cape Fear Garden Club. The Burns-Stokes Garden on Hawthorne Road and Mr. Burns's waterside garden on Banks Channel were tour favorites. This photo was taken at the ribbon cutting in the Thompson garden, in 1982. (Miriam Burns Whitford)

Mrs. Jack C. Thompson, 1938 South Live Oak Parkway

Mr. and Mrs. Robert C. Bauder, 323 South

Front Street

Mr. Bruce DesChamps, 618 Forest Hills Drive

Mr. and Mrs. William J. Stenger, 2210 Marlwood Drive

Mr. and Mrs. Warren Y. Lea, 415 Summer Rest Road

Kenan House, 1701 Market Street

Garvin D. Faulkner, 5 North 20th Street

Dr. and Mrs. John Codington, 624 Forest Hills Drive

Mr. and Mrs. William Echols, 738 Forest Hills Drive

Mr. and Mrs. Richard A. Ryder, 2029 Bradley Drive

Plantation Gardens, Metts Avenue

Mrs. O. R. Hunt chaired the tour that began in the Thompson garden. Roy Clark and Paul Anka were among the guests, but the emerging superstar of the 1982 festival was Michael Jordan. Hugh Morton, North Carolina's photographer-at-large, followed the UNC. Tarheels closely and foresaw Jordan's impending worldwide fame.

Mr. Morton was scheduled to crown the queen, Lynda Goodfriend, an actress who played Lori Beth on "Happy Days." Ten days before the festival he asked if he could "pass the ball" to Michael Jordan. "Michael Jordan, who was a freshman at UNC, sunk the winning basket at the NCAA Tournament," said Mr.

Wilmingtonian Michael Jordan and Hugh Morton crowned Lynda Goodfriend Queen Azalea XXXV, ten days after freshman Jordan sunk the winning shot in the 1982 NCAA Tournament. Here Mr. Morton provides an assist. (Cape Fear Museum)

Morton. "I called and asked the festival officials if they would let Michael crown the queen. They discussed it, then agreed. I handed the crown to him, so I got credit for an assist."

As for garden club activities in 1982, the most important happening was the dedication of the Appleberry Bluebird Trail. Named, of course, for Mrs. Cecil Appleberry, the trail consisted of 29 bird boxes. According to the honoree, "Some of the best birding territory in the U. S. lies within the 15-mile-diameter circle used by the Wilmington Natural Science Club for its annual Audubon Christmas Count."

1983

Mr. Thomas H. Wright III, 312 South Front Street

Mr. and Mrs. Stanley Brooks, 321 South Fourth Street

Dr. and Mrs. Landon Anderson, 520 Orange Street

Dr. and Mrs. John L. Leonard III, 1743 South Live Oak Parkway

The Burns Garden was open for visitors - whether it was on the tour or not. (Photo by Freda Wilkins)

Mr. and Mrs. Carl Brown, 2502 Gillette Drive

Mrs. Peggy Dreyfors, 6217 Richard Bradley Drive

Mr. and Mrs. Warren Barnes, 136 Skystasail Drive

Mr. and Mrs. Ned Dowd, 3 Point Place, Wrightsville Beach

Mr. and Mrs. J. D. Causey, 2037 Shirley Road

Dr. and Mrs. R. V. Fulk, 764 Forest Hills Drive

Mr. and Mrs. Henry B. Rehder, 2217 Oleander Drive

Plantation Gardens, Metts Avenue

Chairman Babs Shaw presented 59 new

The old bridge at Airlie. The Corbett Package Company continued to host the annual barbecue at Airlie until 1999. (Photo by Gilliam K. Horton. Courtesy of Josephine Corbett Horto.)

belles at the 1983 ribbon cutting in the Causey garden. Dressmaker Augusta Counts was honored for her diligence in creating belle gowns. Tina Gayle of the nighttime television series "Chips," was queen. Andy Williams and Barbara Mandrell were featured performers. Proceeds from another successful tour went to St. John's Art Gallery.

1984

Dr. and Mrs. James H. Robinson, 1903 Brookhaven Road

Mr. Joe Gurganus, 418 South Front Street

Deacon Galleries, 109 Castle Street

Appie and Annie Daniels, 1721 Country Club Road

Mr. and Mrs. David Klein, 2010 Graymont Drive

Dr. and Mrs. William Nixon, 1946 South Live Oak Parkway

Mr. and Mrs. Thomas May, 3921 Sweetbriar Road

Mr. and Mrs. Cole Porter, 3962 Appleton Way

The Hon. and Mrs. Gilbert Burnett, River Road

Mr. and Mrs. David Ross, 311 Nun Street

Judy Mowbray chaired the 1984 tour, a labor that produced $11,000. Cape Fear Garden Club distributed the proceeds to meet landscaping needs at the new Bellamy School, Sheltered Workshop, Lower Cape Fear Historical Society, and the Student Activities Building at UNCW.

Queen Susan Wyatt, an actress in the daytime series "All My Children," cut the garden tour ribbon at the home of Dr. and Mrs. James H. Robinson. Singers Barbara Mandrell and Johnny Mathis were performing festival guests.

1985

Mr. and Mrs. Jack Baynes, 2131 Gloucester Place

Drs. William and Catherine Kassens, 2104 South Canterbury Road

Dr. and Mrs. Al Woodworth, 825 Cascade Road

Mr. and Mrs. Eddie Inscoe, Meadowsweet, Greenville Loop Road

Mr. and Mrs. Louis Jordan, Cedar Island

Mr. and Mrs. David Valliere, 6340 Marywood Drive

Mr. and Mrs. Roy Hobbs, 104 Martingale Lane

Mr. and Mrs. Billy Pope, 105 Windlass Drive

Mr. and Mrs. Louis Jordan served up a beautiful Cedar Island landscape for the 1985 tour. (Photo by Freda Wilkins)

Host Peter Browne Ruffin welcomes visitors to the 1986 Cape Fear Garden Club Azalea Festival Tour ribbon cutting. With him are members of his family: Audra Wetherill (belle on left), Suzanne Nash Ruffin, Elise Ruffin, and Ginger Ruffin Wetherill (in pink jacket). (Photo by Freda Wilkins)

The 1985 tour, chaired by Emily McKoy, had some challenges. A fall hurricane (Diana) and a late freeze damaged some scheduled gardens so badly that they were eliminated from the list. Despite that, the tour was hugely successful. Cape Fear Garden Club gave generous checks to the Lower Cape Fear Historical Society and the New Hanover County Arboretum. Phylicia Ayers Allen of "The Bill Cosby Show" reigned supreme over the 1985 festival. Other performing guests included comedian Rich Little, Donny and Marie Osmond, Crystal Gayle, and Alex Trebek, host of NBC's "Jeopardy" show.

Peter Browne Ruffin, 753 Forest Hills Drive

Joseph S. Gurganus, 120 South Third St.

Latimer House 126 S. Third Street (Club project)

Jane and Tom Maloy, 2521 Mimosa Place

Carolyn and Jimmy Fowler, 2308 New Orleans Place

Ann and Bill Grathwol, 2307 New Orleans Place

Fran and Bill Maus, Tannahill, Shandy Avenue

Lee and R. C. Fowler, 349 Shandy Lane

Poplar Grove Plantation, Highway 17, Scott's Hill

Barbara and Henry B. Rehder, 2217 Oleander Drive

Mary Margaret McEachern

"I was such a tomboy during my tenure as an Azalea Belle that I refused to take off my soccer shorts in order to wear the dress. At my mother's insistence, I reluctantly relented, donning the dress, soccer shorts and all. Having to fight the elements in that dress certainly proved a challenge, and I could hardly wait for it to be over so I could get back into my running clothes. As the years go by, however, I recount the experience with more and more fondness. It was a worthwhile once-in-a-lifetime opportunity that I will never forget and would trade for nothing". Mary Margaret McEachern Nunalee

Margaret and Julian H. McKeithan, 1719 Fairway Drive

James Moss Burns, 1417 Hawthorne Road

Dr. Jon H. Gerdes, Jr., 722 Forest Hills Drive

Mrs. P. R. Smith, 615 Forest Hills Drive

Dogwoods canopy a profusion of azaleas in this image of the McKeithan Garden at 1719 Fairway Drive. (Photo by Freda Wilkins)

Peter Browne Ruffin hosted the Queen's garden party in his late wife's garden on Forest Hills Drive. Queen Kim Zimmer played

Kim Zimmer, 1986. (Henry B. Rehder Collection)

"Reva" on the daytime series "Guiding Light." Wayne Newton with his guard dog, Roger Miller, and Barbara Eden of "I Dream of Jeannie" fame were also festival guests. In 1986, Azalea Belles appeared at the festival variety show for the first time.

The Cape Fear Garden Club tour spanned a large area, from the Cape Fear River to Scott's Hill. The tour also featured five gardens that were on the original tour in 1953. Chairman Maryann Robison spoke proudly when she told a reporter, "It's really a nice heritage that we have here in Wilmington with these gardens."

Proceeds from Cape Fear Garden Club tours went to many causes that year: the New Hanover County Arboretum, Latimer House restoration, Cape Fear Technical Institute, the Hinton James building at UNCW, Cape Fear Hospital, Bellamy School, New Hanover County Public Library, the Lucille Shuffler Senior Center, and St. John's Museum of Art.

1987

Marty and Bob King, Greenville Loop Road

Henry J. MacMillan, 118 South Fourth Street

LaDeen and Robert Puddy, 121 South Fourth Street

Jumpin' Run, South 17th Street

Betsy and John Codington, 624 Forest Hills Drive

Hilda and William Echols, 738 Forest Hills Drive

Wanda and Robert Moore, 1936 Brookhaven Road

Martha and Mack Umphlett, 2234 South Live Oak Parkway

Ellen and Dennis Anderson, 2002 Hawthorne Road

Molly and Bill Howell, 222 Masonboro Loop Road

Belles Ashley Greer (standing), Beth Parker Davis (left), and Elise Wilkins (Wilson) formed a pretty trio during the 1987 tour. (Photo by Freda Wilkins)

Tour chairman Mary Lou McEachern and hundreds of other volunteers helped raise $17,000 with the 1987 tour. After the ribbon was cut at the King garden, Cape Fear Garden Club's hospitality committee served punch and cake to 1000 guests. Beth Chadwick Cherry, one of the original azalea belles, chaired the belle committee and helped advise 69 teenage girls.

Queen Robin Greer, who played the role of an aerobics instructor on the CBS nighttime drama "Falcon Crest," opened the garden tour at the 13-acre estate of Mr. and Mrs. Bob King. The Kings property has now been subdivided into an area known as White Hall, but the main house has been used numerous times in movies and in television shows like Andy Griffith's "Matlock."

Robin Greer, 1988 (Henry B. Rehder Collection)

Actor Lorenzo Lamas, described by Wilmington writer Celia Rivenbark as the "hunk in residence" at "Falcon Crest," was also a guest. In addition to the duo from "Falcon Crest" appearing together at the 1987 festival, the show had a number of direct and indirect ties to the Azalea Festival. Ironically, Lamas married Kathleen Kinmont, daughter of his onscreen "Falcon Crest" mother, Abby Dalton, who was azalea queen in 1964. And in turn, Jane Wyman, Ronald Reagan's first wife, played Abby Dalton's mother on "Falcon Crest."

1988

Garden club member Rena Nelson MacRae (second from right), a tree lover, was pleased that her husband, Hugh MacRae, donated 101 acres of land to New Hanover County as a perpetual pine forest. Others present at the 1954 dedication were county commissioner Claude O'Shields, Agnes MacRae Morton (Hugh MacRae's daughter), and commissioner Ralph Horton.(Photo by Hugh Morton, Cape Fear Museum

Elizabeth and Thomas Wright, 2232 Acacia Drive

Janice and Jack Dunn, 2901 Oleander Drive

Sherron and Dan Moore, 1302 HawthorneRoad

Ann and John Parker, 3821 Gillette Drive

Pond Gardens at Hugh MacRae Park

Jo Ann and William Stenger, 2210 Marlwood Drive

Saundra and David Sikes, 4629 Carolina Beach Road

Betty and Claude Hill, 4810 Wilderness Road

Lynda and James Roesch, 706 Jacobs Creek Lane

New Hanover County Extension Service Arboretum, 6206 Oleander Drive

Pamela and Russell Grass, 2309 New Orleans Place

Judi and Nathan Sanders, Summers Rest Road

Hugh MacRae Park in the 1950s. The little boy is Hugh Morton, Jr. (Photo by Hugh MacRae Morton, Cape Fear Museum)

the Cape Fear Garden Club. Certainly she influenced her husband in his decision to give 101 acres of pine forest to New Hanover County. Then his grandson, Hugh Morton, helped organize the Azalea Festival —and eventually the Cape Fear Garden Club included the park on the annual festival tour.

Sometimes Cape Fear Garden Club appends a free attraction to the tour, like the Pond Gardens at Hugh MacRae Park. Invariably locals discover a new favorite haunt "in their own back yard." (Photo by Freda Wilkins)

The inclusion of the Pond Gardens at Hugh MacRae Park on the 1988 garden club tour seemed to close a certain circle. In 1925, Mrs. Hugh MacRae was a founding member of

"I was with my grandfather at James Walker Hospital the day before he died," said Hugh Morton, in 2003. "'You take care of that park,' he told me. And I did. At one time, there was a campaign to locate St. Andrew's College in Wilmington, but the deal hinged on the school's acquisition of Hugh MacRae Park. I fought it and St. Andrew's was built in Laurinburg. Later, New Hanover County asked to build a school there and promised to name it Hugh MacRae High School. I refused and they moved across the highway and built John T. Hoggard High School."

The 1988 tour, chaired by Jo Pope, made $18,000 that was rerouted to the community

The Pond Gardens at Hugh MacRae Park. (Photo by Freda Wilkins)

for beautifications projects. The shade garden at the New Hanover County Arboretum was dedicated to Cape Fear Garden Club in honor of a whopping $30,000 in contributions from various club projects to the rapidly evolving garden showplace.

Kim Morgan Greene, a veteran of daytime dramas, was Queen Azalea XLI. Since 1988, she has played guest starring roles on scores of prime-time shows. Singers Tom Jones and Neil Sedaka also appeared at the 1988 festival.

It was the crowd that roared when singer Tom Jones launched into his golden oldie, "What's New Pussycat" during the 1988 festival. (Photo by Edwina Batson)

Meg Talbert and Steve Davenport, 741 Forest Hills Drive

St. James Church Memorial Garden, 25 South Third Street

Orrell and David Alan Jones, 117 South Fourth Street

Mary Lee and W. K. Stewart, 2203 Marlwood Drive

Winifred and Blackwell B. Pierce, 2312 Blythe Road

Mary Jo and Robert Hundley, 1840 Hawthorne Road

Betty and Ed Rusher, 1322 Country Club Road

Elizabeth and Oscar Waldkirch, 5303 Greenville Loop Road

Mable and Gilbert Parrish, Greenville Loop Road

New Hanover County Extension Service Arboretum, 6206 Oleander Drive

By 1989, the Azalea Festival parade had grown from 16 floats to almost 200, a fact that must have delighted parade marshal Hugh Morton. Celebrities led the parade, a group that included Jed Allen of the TV show "Santa Barbara" and "Dallas" actress Beth Toussaint.

Queen Azalea XLII, actress Rebecca Arthur of the television show "Perfect

Lady Banksia roses help frame this beautiful scene at the Rehder Garden. Though Cape Fear Garden Club members can appear to be demure adult Southern belles, they are actually moneymaking powerhouses that channel their organization's profits back into the community. (Photo by Freda Wilkins)

Strangers," opened the Cape Fear Garden Club tour at the home of Meg and Steve Davenport on Forest Hills Drive. The Davenports purchased the home of Mr. S. L. Marbury, and found themselves the owners of a veritable camellia farm. They manicured what already existed and brought in new plants as well. Despite damaging rains that fell before the festival, the Davenport garden was pristine, delighting guests and thrilling tour chairman Sandy May.

Festival entertainment included five stages at the street fair that featured a wide range of music. Strollers could listen to their pick of music including the Dixieland Society of the Lower Cape Fear, Eve Cornelius, and the UNCW Gospel Choir. Frankie Valli and the Four Seasons were also highlights at the festival.

1990

Mr. and Mrs. George W. Jones, Jr., 510 Surry Street

Mr. Joseph S. Gurganus, 120 South Third Street

Mr. and Mrs. Henry B. Rehder, 2217 Oleander Drive

Mr. and Mrs. Julian M. McKeithan, 1719 Fairway Drive

Dr. and Mrs. S. Clayton Callaway, Jr., 2312 Gillette Drive

Mr. and Mrs. D. Webster Trask, 2111 South Churchill Drive

Mr. and Mrs. William T. Smith III, 2802 Park Avenue

Mr. and Mrs. Fred Willetts, Jr., 1103 Windsor Drive

Dr. and Mrs. Bertram Williams, 1114 Forest Hills Drive

Mrs. M. G. Allison, Middle Sound Road

Mr. and Mrs. Jabe V. Hardee, 409 Highgreen Drive

The 1990 garden tour began on Surry Street at the George Jones residence, a house built about 1800 that was moved 50 yards west in 1973 to the spot where the Broadfoot Iron Works was once located. Another favorite spot,

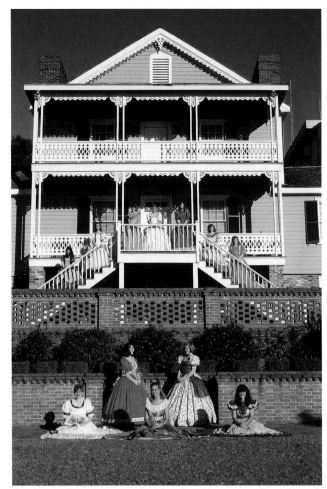

Azalea belles dress up the riverfront house and garden of George Jones, Jr., at 510 Surry Street. (Photo by Freda Wilkins)

the McKeithan garden, was featured in 1986 in *Grand Homes of the South and North Carolina.* A mass of pink azaleas and a canopy of dogwoods made it a wonderland at night, when the garden was lighted for visitors.

Two gardens on the 1990 tour, chaired by Grace Hobbs, were actually double features: the Smith Garden on Park Avenue included the yard of the next-door neighbor, Bill Harris. Mr. and Mrs. John J. Burney teamed up with their neighbors, Dr. and Mrs. R. Bertram Williams.

The 1990 Queen was Kate Collins, a daytime drama actress. Musical guests Kenny Rogers and the group Alabama also provided festival entertainment.

Queen Kate Collins cuts the ribbon for the 1990 Cape Fear Garden Club Azalea Festival Tour. Elma Porter Bowden (on right) is a longtime Cape Fear Garden Club volunteer.(Photo by Edwina Batson)

Mr. and Mrs. Robert C. Bauder, 323 South Front Street

Mr. and Mrs. Allen Bech, 206 Orange Street

Mr. and Mrs. Steven Harper, 2221 Acacia Drive

Mr. and Mrs. Waylon Lynch, 417 Colonial Drive

Mr. and Mrs. Clyde Clark, 3835 Sylvan Road

Mr. and Mrs. Milton Fuller, 3937 Halifax Road

Mr. and Mrs. A. W. Stafford, 401 Bradley Creek Point Road

Mr. and Mrs. James T. Balkcum, Jr., 156 Edgewater Lane

Dr. and Mrs. Donald D. Getz, 1101 Airlie Road

Mr. and Mrs. Kirk Crumpler, 7225 Gray Gables Road

Mr. S. N. McKenzie, 4 Island Drive

One of the many interesting features about the 1991 tour, chaired by Naomi Causey and Anne Redwine, was the garden of Mr. and Mrs. Kirk Crumpler at Gray Gables. The immediate former home of the Pearsalls, it featured a formal English garden thought to have its origins in the 19th Century. The Crumplers restored statuary, fountains, walls, and benches

Effie Barefoot Burney (on left) and Billie Snipes Hinson were beloved, long-time veterans of garden tour duty when this picture was taken in 1991. (Photo by Dianne Lynch)

to create a sort of garden museum. A slave house dating to 1849 made a thought-provoking site.

The Lynches' Garden also included an interesting feature. Owners Dianne and Waylon Lynch created an American flag in flowers as a symbol of support for U. S. troops in the Gulf War.

Queen Tonya Walker opened the 1991 tour at Edgewater, the Getz home at 1101 Airlie Road. A wistful belle (on left) and belle gown designer, Alma Fennell, support her. (Photo by Edwina Batson)

Tonya Walker of the daytime drama "One Life to Live," was Queen Azalea in 1991. Bill Cosby performed before an audience of 6,000 people, in

Bill Cosby performs at Trask Coliseum in 1991. (Photo by Edwina Batson)

UNCW's Trask Coliseum, and The Judds put in one of their last appearances together. "Match Game" host Ross Shafer, "Knots Landing" star Lorenzo Caccialanza, and actor Richard Kind were also festival guests.

1992

Dr. and Mrs. John Codington, 624 Forest Hills Drive

John Crowley and Richard Tomes, 121 South Second Street

Dr. and Mrs. Robert Kelly, 2309 Knightsbridge Road

Mrs. Annie Gainey, 3140 Kirby Smith Drive

Mr. and Mrs. Lawrence Sanders, 105 Chimney Lane

Mr. and Mrs. Andy Young, 2717 Shandy Lane

Mrs. Peggy Dreyfors, 511 Bradley Creek Point Road

Mr. and Mrs. Paul Moyle, 525 South Lumina Avenue

Mr. and Mrs. Willis Brown, 1714 Landfall Drive

Mr. and Mrs. Richard Hopper, 916 Twisted Oak Place

Anne Redwine chaired the festival tour and Betty Hill orchestrated the Azalea Belles in 1992. Proceeds from ticket sales went to a wide variety of projects including three New

Hanover County agencies and St. John's Museum of Art.

Clare Carey, Azalea Queen XLV, played Kelly Fox, Hayden Fox's daughter in the ABC series "Coach." Ironically, Shelley Fabares, the 1961 queen, starred in the same series. During the festivities, Ms. Carey enjoyed the company of fellow television stars Amy Aquino of "Brooklyn Bridge" and J. D. Roth of "Fun House."

Attended by the late Johanna Rehder (on left) and Meg Gemmell (Sperry), Kelly Ripa cut the ribbon at the Kenan House in 1993. (Photo by Elaine Henson)

Queen Clare Carey credited her dance training when she balanced atop a police motorcycle in 1992. (Photo by R. T. Brown of Perkins Gallery, Henry B. Rehder Collection)

Adding her own distinct signature to the event, Clare Carey posed in orange for the customary photo on the back of a police motorcycle. The result is a perpetual testimony to some vital elements of the festival: vivid color, natural beauty, and local government's necessary participation.

Kenan House, 1705 Market Street

Dr. Lance Wright, 217 South Second Street

Dr. and Mrs. Landon Anderson, 520 Orange Street

Mr. and Mrs. Vincent A. Romano, 6609 River Road

Mr. and Mrs. Domenick Rella, 4011 Chapra Drive

Mr. and Mrs. William Howard and Mrs. Beatrice Young, 4000 Chapra Drive

Mr. and Mrs. W. J. Blair, Jr., 1900 South Live Oak Parkway

Mr. and Mrs. R. C. Fowler, 2708 Shandy Lane

Capt. and Mrs. Richard Ryder, 229 Bradley Drive

Dr. and Mrs. Charles M. Hicks, 9308 New Orleans Drive

Mr. and Mrs. W. P. Sineath, 6664 Cable Car Lane

Mr. and Mrs. William Wright, 1063 Ocean Ridge Drive

Mr. and Mrs. Paul Mullan, 2345 Ocean Point Drive

Mr. and Mrs. John Young, 24 Shore Drive

Tour chairman Alma Fennell created a diverse list of gardens for the 1993 festival. Guests could wander through the stately Kenan House garden, as well as new spaces at Landfall and Wilmington's booming subdivisions near Wrightsville and Greenville Sound. Historic district gardens added to the mix and showed tour patrons what could be cultivated in a limited amount of space.

Queen Kelly Ripa didn't seem to mind being the only girl keeping company with the Citadel Summerall Guards, April 2, 1993. A crack precision team, the cadets perform in front of City Hall every year on parade day, and gather second glances most everywhere else as well. (Photo by Edwina Batson)

Kelly Ripa, of "All My Children" was Queen Azalea XLVI. Ms. Ripa, the former

queen who currently has the most recognizable name, stars with Regis Philbin weekday mornings on the "Regis and Kelly Show." She still has a part in "All My Children" and is also planning a prime-time role.

1994

The Burgwin-Wright garden. (Photo by Barbara Marcoft, Cape Fear Museum)

Woodbury Garden, 721 Forest Hills Drive

Echols Garden, 738 Forest Hills Drive

Rehder Garden, 2217 Oleander Drive

Ligon B. Flynn Architects, 15 South Second Street

Burgwin-Wright House, 224 Market Street

Lees Garden, 906 Robert E. Lee Drive

Bush and Blick Gardens, 4433 and 4429 Windtree Road

Harold Garden, 1713 Signature Place

Dunn Garden, 1235 Great Oaks Drive

The Lees garden on Robert E. Lee Drive provided variety to the 1994 tour, chaired by Pat Geyer. A formal Japanese garden with stone seats and a pond, the space seemed entirely appropriate in light of Mrs. Lees's vocation. Ruth Lee and Bonnie Burney, daughter-in-law of longtime Cape Fear Garden Club member Effie Barefoot Burney, own Ikebana, a florist strong on Japanese-style arrangements.

The entertainment in 1994 was diverse. Though Frank Sinatra suffered a fall just 12 days before the festival, he appeared and sang for thousands of fans. Country singer Ronnie Milsap sang, as did Frankie Valli and the Four Seasons. The Clyde Beatty-Cole Brothers

Laura Sisk and Henry Rehder, 1994 (Henry B. Rehder Collection)

Circus and the Azalea Festival Street Fair rounded out the offerings.

Laura Sisk, who played Allison Rescott on the ABC daytime drama "Loving" was Queen Azalea XLVII.

1995

A rare yellow azalea blooms in the garden of LaVonne and Ken Gault. (Photo by Freda Wilkins)

Naomi and J. D. Causey, 2037 Shirley Road

Virginia and Jim Pierce, 1151 Forest Hills Drive

Barbara and Henry Rehder, 2217 Oleander Drive

Betty and James Hooten, 2110 South Canterbury Road

Cathy and Lester Wilson, 5602 Lands End Court

Sally and Frank Crandall, 5625 Green Turtle Lane

Russell LaBelle, 503 Bradley Creek Point Road

Jean and Calvin Ross, 1305 Bar Harbor

Dixey and Dick Smith, 2000 Marsh Harbor Place

Gail and Wilbur Tice, 1247 Great Oaks Drive

Queen Laura Bonarrigo and Henry Rehder. (Henry B. Rehder Collection)

Queen Laura Bonarrigo, who played journalist Cassie Callison on "One Life to Live," cut the ribbon at the home of Naomi and J. D. Causey. In addition to some of the better-known gardens, the tour, lined up by chairman Carolyn Augustine, also included the Crandall home, a low-lying garden on the lake in Turtle Hall and the LaBelle garden on Bradley Creek Point.

Turtle Hall, on Greenville Sound, was once an estate that belonged to Daisy and Rye Page in the early days of the garden Pilgrimage. Developed in the 1980s, the subdivision still contains some exotic fruit trees and plants established there by Daisy Page, a world traveler who loved to garden.

Hugh MacRae Park. (Photo by Freda Wilkins)

Bradley Creek Point, originally known as Mount Lebanon, still boasts some of the cedar trees for which the land got its Biblical name. Though originally both Turtle Hall and Bradley Creek Point had main houses purposefully located on precipices far from the high water line, both are now developed to the water's edge, riskier sites in great storms but ones that provide beautiful backdrops to colorful spring gardens.

1996

Lee and Nick Garrett, 6310 Greenville Sound Road

Lee and R. C. Fowler, 2708 Shandy Lane

Betty and John Burney, 1130 Forest Hills Drive

Margaret and Julian McKeithan, 1719 Fairway Drive

Linda and Gene Renzaglia, 1310 Spotswood Court

Alice and Ron Sullivan, 130 Bradley Pines Drive

Kemp Burpeau, 105 Parmele Drive

Leslie and Jim Hively, 2005 Spanish Wells Drive

Peggy and John Yelverton, 1019 Ocean Ridge Drive

Gerry and Frank Ceravolo, 813 Swift Wind Place

Jenene Smith chaired the 1996 garden tour, an effort that raised $38,769 for such worthy community efforts as Lower Cape Fear Hospice, Hospitality House, and the Bellamy Mansion.

Queen Gina Tognoni, an actress who portrayed Kelly Cramer Buchanan in the day-time drama "One Life to Live" opened the tour at the Garretts' home on Greenville Sound. Ms. Tognoni was a former Miss Teen All America, a title shared by Tonya Walker, Azalea Festival queen in 1991, and superstar Halle Berry.

Willie Stargell served as festival grand marshal. A former baseman and outfielder for the Pittsburg Pirates, he led his team to two World Series titles, in 1971 and 1979. The baseball star married Wilmingtonian Margaret Weller, twin sister of local newscaster Frances Weller. Sadly, Mr. Stargell died August 9, 2001, at the age of 61.

1997

Temple of Love, Landfall Drive

Rosenberg Garden, 1924 South Live Oak Parkway

Rehder Garden, 2217 Oleander Drive

Simmons Garden, 1215 Windsor Drive

Eason Garden, 711 Forest Hills Drive

Wise Alumni House Garden, 1713 Market Street

Kenan House Garden, 1705 Market Street

Jones Garden, 1417 Market Street

Bellamy Mansion Garden, Fifth Avenue and Market Street

Sullivan-Murchison Garden, 107 South Fourth Street

Campbell Garden, 118 South Fourth Street

The Temple of Love at Landfall, designed by Jefferson Memorial architect John Russell Pope, took center stage at the Queen's Garden Party. Though the original pools that surrounded it had been destroyed by developers, new ground had been cultivated to embellish the coquino gazebo.

The Bellamy Mansion Garden returned to the tour, chaired by Shirley Hardee, after efforts were made to recreate the original plan.

Still at it: Two festival organizers, Kenneth Sprunt and Hugh Morton, chat with Miss North Carolina 1997, Michelle Warren, following Mr. Morton's inclusion in the Walk of Fame. (Photo by Susan Block)

Lauren Roman. (Henry B. Rehder Collection)

Some of the earlier plants had been brought from Germany by Mrs. Henry Rehder, Sr.

Queen Lauren Roman, raised in Virginia, returned to her birthplace when she arrived for the festival. Born in 1975, the young queen already had a part in a daytime drama. To date, she has starred in "All My Children," "The Bold and the Beautiful," and has played a role in "Buffy the Vampire Slayer."

Miss Black USA Matilda Caroline Mack was a 1997 festival guest. Vince Gill, Frankie Valli, and The Four Tops entertained at Trask Coliseum.

During the 50th Azalea Festival, the city of Wilmington dedicated a star on the riverfront Walk of Fame to 1997 Grand Marshal Hugh Morton. A glowing tribute was read to celebrate his many achievements, including his leading role in getting the festival started and keeping it going during the early years. A true-blue Carolina Tarheel and thus a team player, Mr. Morton took the podium and passed the credit along to Dr. Houston Moore.

1998

Parsley Estate, 7527 Masonboro Sound Road

Plaskett Garden, 6603 Cove Point Drive

Grimes Garden, 6216 Stonebridge Road

Rodger Garden, 905 Rabbit Run

Gay Garden, 2700 Shandy Lane

Fuller Garden, 106 Martingale Lane

Moore Garden, 623 Colonial Drive

Todd Garden, 219 Keaton Avenue

Howard Garden, 4000 Chapra Drive

Mees Garden, 233 Devonshire Lane

Leigh Hobbs Murray was chairman of the 1998 Cape Fear Garden Club tour, an assemblage that included gardens on Masonboro and Greenville Sound. Live Oaks, the 1912 house designed for Agnes MacRae and Walter Parsley by Lincoln Memorial architect Henry Bacon, was the centerpiece of one of the year's most popular gardens.

Queen Alla Korot was born in what is now Ukraine in 1970. Although she was known primarily in 1998 as a daytime drama actress, having had roles in "All My Children" and "Another World," she has now had many guest starring roles in nighttime dramas as well. Her most notable recent appearances have been as Erin Vratalov in the prime-time series "The District." Additionally, she had parts in several motion pictures: "Night of the Cyclone," The Colony," "Shoo Fly," and "Free."

1999

Mrs. Wilbur Corbett, 1310 Country Club Road

Mr. and Mrs. David Kincade, 2416 Mimosa Place

Mr. and Mrs. James Nichols, 336 West Renovah Circle

Mr. and Mrs. Vance Young, 1052 Ocean Ridge Drive

Mr. and Mrs. Dave Phillips, 1035 Ocean Ridge Drive

Mr. and Mrs. Tom Morris, 1016 Arboretum Drive

Mr. and Mrs. Jack Weymouth, 1312 Pembroke Jones Drive

Mr. and Mrs. John Ferebee, 917 South Lumina Avenue

Mr. and Mrs. Dennis Gillings, 10 East Oxford Street

Mr. and Mrs. Larry Clifton, 2601 North Lumina Avenue

Alla Korot in the Rehder garden. (Photo by Edwina Batson)

The old pergola fountain at Airlie, about 1948. In 1999, New Hanover County purchased the garden. (Photo by Gilliam K. Horton. Courtesy of Josephine Corbett Horton)

Airlie's famous black swans and cygnets. (Photo by Gilliam K. Horton. Courtesy of Josephine Corbett Horton)

Joanne Corbett, a Fulbright scholar and former professor of English at UNC-W, had always had a pretty garden. Landscaped originally by Jim Ferger, it framed the house well. But during the 1990s, when Dr. Corbett decided she wanted to spend more time outside, she employed Shaw Burney to design some embellishments. Assisted by Dr. Corbett's daughter, Melissa, Mr. Burney created formal spaces and beautiful borders.

So when Shannon Corbett Maus, tour chairman in 1999, couldn't find a garden owner to host the tour opening, she called on her mother. Dr. Corbett agreed and the result was an especially pretty scene, worthy of a family with ties to Airlie. Dr. Corbett and her husband, Wilbur, had lived one summer in the rambling old mansion on Bradley Creek.

Sydney Penny, who played Julia Santos Keefer in the daytime drama "All My Children," was Queen Azalea LII. Astronaut

The Airlie pergola. (Photo by Gilliam K. Horton. Courtesy of Josephine Corbett Horton)

Curtis L. Brown, Jr., of Elizabethtown was Grand Marshal of the 1999 festival parade.

2000

The Longley Garden. (Photo by Elaine Henson)

Dr. and Mrs. Tom Maloy, 2521 Mimosa Place

Mr. and Mrs. Henry Longley, 2525 Mimosa Place

Mr. Julian McKeithan, 1719 Fairway Drive

The Julian McKeithan garden. (Courtesy of Elaine Henson)

Mr. and Mrs. J. D. Causey, 2037 Shirley Road

Mr. and Mrs. Donald Christopher, 2209 Marlwood Drive

Mr. and Mrs. Tom Spivey, 1700 Softwind Way

Mr. and Mrs. Dennis Moeller, 1701 Softwind Way

Mr. and Mrs. Alan Jones, 3009 Rachel Place

Mr. and Mrs. William Wright, 1063 Ocean Ridge Drive

Mr. and Mrs. Alan Bede, 1216 Great Oaks Drive

In 2000, the 75th anniversary of Cape Fear Garden Club, Jo Chadwick retired as longtime treasurer of the Azalea Festival tours. She had seen tabulations run from a few thousand to $40,000 as the tour years progressed.

Anna Goolsby (Toconis) led the garden club tour in 2000. The list included some old favorites as well as new gardens at Landfall, always a popular venue for guests.

Nina Repeta, a cast member of the Wilmington-based television show "Dawson's Creek" was Queen Azalea LIII. Ms. Repeta, a Shelby native, also played in the movie, "The Angel Doll."

2001

The 2001 ribbon cutting took place at the home of Patty and John Baker. (Photo by Freda Wilkins, Courtesy of Elaine Henson)

Patti and John Baker, 217 Colonial Drive

Linda and Warren McMurry, 218 Forest Hills Drive

Katherine and Marty Baker, 2222 Metts Avenue

Jo Robinson and Donald Suder, 502 Decatur Drive

New Hanover County Arboretum, 6206 Oleander Drive

Dot and Jim Balkcum, 156 Edgewater Drive

Gail Tice, 1721 Fontenay Place

Beloved tour assistant Percy Glaspie became an honorary member of Cape Fear Garden Club in 2001. Mary Lou McEachern (on right) has been garden club tour and Azalea Belle sponsor for many years. (Photo by Millie Maready)

Carol and Ned Olds, 1047 Ocean Ridge Drive

Janice and Reuben Allen, 809 Oyster Landing

Mary Jo and Gary Shipman, 6512 Carmel Trail

Tannahill, the Greenville Sound home of Shannon Corbett and Bill Maus, created a beautiful setting for six Azalea Belles in 2001. The girls are: (front row, left to right) Jessica Kesler, Erin Anderson, Catherine Gerdes, Louise Rippy, (back row, left to right) Amy Davis and Maggie Mowbray. (Photo by David Wittmer)

Dianne Lynch, who chaired the 2001 tour, took delight in the fact that the two belles chosen to escort the queen were named Tara and Ashley.

Nikki DeLoach, 2001. (Henry B. Rehder Collection)

Lots of locals cheered when veteran tour assistant Percy Glaspie was voted in as an honorary member of Cape Fear Garden Club.

Nikki DeLoach, lead singer of Innosense, a five-woman band, was queen in 2001. A former Mouseketeer on "The New Mickey Mouse Show," she met singers Justin Timberlake and Britney Spears. Innosense toured with Timberlake's band 'N Sync in 2000.

Grammy winner Tony Bennett performed in Trask Coliseum and sang his signature tune, "I Left My Heart in San Francisco."

Henry Rehder between 2 Azalea Queens: Nikki DeLoach (left) and Nina Repeta (2000 queen) (Photo by Edwina Batson)

Festival volunteer Gayle Ward snapped this photo of her beloved belles in 2002. For years, Mrs. Ward led a belle training course. "No hoop earrings. No athletic shoes. No Walkmans."

2002

Dorothy and Terry Mildenberg, 6013 Wellesley Drive

Alison and Don Getz, 6233 Tortoise Lane

Marie and James Cooke, 6432 Westport Drive

Denise and Dan Smith, 1255 Great Oaks Drive

Becky and Bill Salter, 1235 Great Oaks Drive

Chris Lindley, 3702 Wrightsville Avenue

Aggie and Jack Henriksen, 325 East Renovah Circle

Eleanor and Thomas Hissam, 2601 Park Avenue

Martha Beery, 1919 Brookhaven Drive

Goldie and Mike Stetton, 7518 Masonboro Sound Road

Nancy and David Kauffman, 7500 Jonquil Court

Teresa Hill chaired the 2002 tour, a roster of gardens that covered everything from a cottage garden, to a Bali garden, to the intact terraces that once bordered Pembroke Jones's famous hunting lodge that he called "The Bungalow," at what is now Landfall. Proceeds from the tour, totalling at $52,000, were distributed among various local projects.

Teresa Hill, who chaired the 2002 Cape Fear Garden Club Azalea Tour, has been an advocate for the chronicling of the organization's achievements. (Photo by Elaine Henson)

Queen Valerie Wildman, better known as an actress in "Days of our Lives," was also a former Peace Corps volunteer and a spokesperson for children's issues for the National Coalition. Elizabeth Hanford Dole, a Salisbury native and a candidate for the U. S. Senate, was a special festival guest in 2002. It was actually a return visit: In 1958, as the Duke University May Queen, she was a member of the queen's court.

WECT-TV (NBC) personality Bob Townsend has been a good friend to Cape Fear Garden Club, providing spirited coverage of Azalea Garden Tour openings for years. He is pictured here in 2002, interviewing Queen Valerie Wildman.

Profits from the 2002 tour were disbursed among 11 organizations: Residents of Old Wilmington, Airlie, Ability Garden, Bellamy Mansion, Latimer House, Winter Park Elementary School, Cameron Art Museum, Kids Making it, Pine Valley PTA, Cape Fear Community Foundation, and Roland-Grise PTA.

Mount Lebanon Chapel, built in 1835, was probably the earliest Gothic structure built in the Wilmington area. Though sequestered in Airlie, it is owned by St. James Church and is the oldest house of worship in the county. (Photo by Gilliam K. Horton, about 1948. Courtesy of Josephine Corbett Horton)

2003: *A Golden Anniversary*

Lillian and Percy Smith, 615 Forest Hills Drive

Lorraine and Alan Perry, 2304 Metts Avenue

Diana and Dennis Overton, 722 Forest Hills Drive

Henry B. Rehder, 2217 Oleander Drive

Mimi Burns and Rick Whitford, 1417 Hawthorne Road

Bellamy Mansion, 503 Market Street – Beverly Ayscue, Director

Elaine and Larry Neuwirth, 300 South Front Street

Pierce and Bill Overman, 312 South Front Street

Airlie Gardens and Mount Lebanon Chapel, 300 Airlie Road - Thomas Herrera-Mishler, director and The Rev. Ron Abrams, rector of St. James Church

Jackie and William Warwick, 2004 Balmoral Place

Ann and Anthony Lees, 6410 Timber Creek Lane

Azalea Belle Margaret Williams and 2003 Azalea Garden Tour chairman Elaine Henson at the Overton Garden. (Photo by Charles V. Henson)

Elaine Henson, chairman of the 50th Anniversary Tour, created a garden pilgrimage that showcased many properties on the 1953 tour. Mrs. Smith's garden was in the hands of her older son, Percy, and his wife, Lillian. Lorraine and Alan Perry had lovingly cultivated the old Millican garden. Diana and Dennis Overton had recently purchased the Gerdes

Azalea Tour volunteers Cathy Poulos, Goldie Stetten, Elaine Henson and Brenda Moore enjoy the Queen's Welcome at Greenfield Lake. (Courtesy of Elaine Henson)

Elaine Henson, Percy Smith, Mary Lou McEachern, and Dot Bryant pose at the 50th anniversary garden tour. (Photo by Freda Wilkins)

Azalea Belle Kara Greer and tour chairman Elaine Henson. (Photo by Martha Greer)

State garden club president Jane Barbot, Cape Fear Garden Club volunteers Anne O'Malley (left) and Shirlee Hardee enjoy a colorful moment at the 2003 ribbon cutting. (Photo by Freda Wilkins)

property. The Bellamy Mansion and Airlie Gardens had moved from private hands to public ones. Three of the original gardens were still owned by the original families: the Rehders, Smiths, and Burns.

The 50th Anniversary. Percy and Lillian Smith greet visitors to the same garden Percy's mother, Bess, opened for the first Azalea Tour in 1953. (Photo by Freda Wilkins)

Local president Dianne Lynch and state president Jane Bardot. (Photo by Freda Wilkins)

A portal of the Smith's "garage" frames the 2003 ribbon cutting, 50 years after the tour was proposed on the very same spot. (Photo by Freda Wilkins)

Jenene Smith has long been a champion of the Azalea Belle program and a host of other Cape Fear Garden Club programs. (Photo by Freda Wilkins)

Queen Tracey Bregman, an Emmy award-winning actress, cut the ribbon at the Smith garden. Ms. Bregman starred as Lauren on the CBS daytime drama "The Young and The Restless."

Young Mimiam Burns (Whitford) enjoys her grandmother's garden, about 1954. (Courtesy of Mimi and Rick Whitford)

Since 1948, the North Carolina Azalea Festival Committee, now consisting of more than 1000 volunteers, has carried on the tradition of showcasing our beautiful Wilmington community while making a significant economic impact annually to our region. The Cape Fear Garden Club has been a vital part of carrying on the Azalea Festival legacy by organizing tours of treasured Wilmington gardens for the past 50 years. It is fitting that a book has been written to capture the beauty of the local gardens, and the grandeur of the Azalea Festival.

On behalf on the entire North Carolina Azalea Festival, I would like to thank the Cape Fear Garden Club for assisting Susan Block in this project. This book is guaranteed to be a keepsake for all who love Wilmington, our rich heritage, the Azalea Festival, and of course, our spectacular gardens. Finally, we express our sincere appreciation to Susan Block for preserving a written history of the Cape Fear Garden Club and the North Carolina Azalea Festival.

Heaven's Belles. (Photo by Freda Wilkins)

Lisa M. Ballantine
2004 President
North Carolina Azalea Festival

Cape Fear Garden Club Presidents:

PAST PRESIDENTS
1951 - 1954

MISS ALLIE MORRIS FECHTIG
JANUARY-MAY, 1951

MRS. P. R. SMITH
1951-1952
1953-JANUARY, 1954
(Resigned January 6, 1954)

MRS. U. LEE SPENCE
1952-1953

MRS. HUGH MORTON
JANUARY-MAY, 1954

(New Hanover County Public Library)

Mrs. N. M. Martin
(Organizing President)
1925-27
Mrs. J. H. Hamilton
1927-29
Mrs. J. B. Cranmer
1929-1931
Mrs. James Sprunt Hall
1931-1932
Mrs. William Latimer
1932-1934
Mrs. R. H. Hubbard
1934-1936
Mrs. C. D. Maffitt
1936-1938
Miss Allie Morris Fechtig
1938-1941
Mrs. A. H. Elliot
1941-1944
Mrs. J. Henry Gerdes
1944-1945
Mrs. Daisy Page Hutaff
1947-1949
Miss Allie Morris Fechtig
1949-1951
Mrs. P. R. Smith
1951-1952
Mrs. U. Lee Spence, Jr.
1952-1953
Mrs. P. R. Smith
1953-1954
Mrs. Hugh Morton
1954
Mrs. Andrew H. Harriss, Jr.
1954-1956

Mrs. Roger C. McCarl
1956-1958
Mrs. James W. Lamberson
1958-1959
Mrs. E. M. McEachern
1959-1961
Mrs. R. C. Andrews
1961-1963
Mrs. A. W. Blount
1963-1965
Mrs. Charles J. Blake
1965-1967
Mrs. Harley E. Vance
1967-1969
Mrs. Allan D. Howland
1969-1971
Mrs. W. K. Stewart
1971-1973
Mrs. Conrad Schwartz
1973-1975
Mrs. O. Raymond Hunt
1975-1977
Mrs. Bruce A. Bryant
1977-1979
Mrs. James J. Pence
1979-1981
Mrs. Lucien Wilkins
1981-1983

Mrs. Donald Christopher
1983-1985
Mrs. James C. Barker
1985-1987
Mrs. Richard F. Flynn
1987-1988
Mrs. Elma Porter Bowden
1988-1990
Mrs. Donald Fennell
1990-1991
Mrs. William Pope
1991-1992
Mrs. Jack Newton
1992-1993
Mrs. Clayton B. Smith, Jr.
1993-1994
Mrs. Joseph C. Knox, Jr.
1994-1995
Mrs. Glenn Avery
1995-1996
Mrs. Bill Huffine
1996-1997
Mrs. Joseph Augustine
1997-1998
Mrs. Lilmar Taylor-Williams
1998-1999
Mrs. Leigh Hobbs Murray
1999-2000
Mrs. Shirley Hardee
2000-2001
Mrs. MaeOmie Mosely
2001-2002
Mrs. Dianne Lynch
2002-2003

MAKING IT OFFICIAL—Dr. W. Houston Moore, who conceived the idea of an Azalea Festival for Wilmington, signs the certificate of incorporation for the Festival, as other members of the general committee, who also signed the sertificate, look on. In the picture are 15 of the 27 persons who signed the certificate at a meeting of the committee in the Woodrow Wilson hut last night. Standing behind Dr. Moore are, left to right: Mrs. George L. Dicksey, Miss Verna Sheppard, Mrs. W. A. Fonvielle, C. F. Taylor, Joseph J. Ray, H. L. McPherson, L. C. LeGwin, Jr. James R. Benson, C. C. Johnson, Billy Burns, Stanley Rehder, John E. Hope, and W. H. Corbett. Seated at the end of the table is Kenneth M. Sprunt, treasurer of the Festival.

—Photo by Hugh Morton.

(*Star News* photograph, courtesy of Betsy and Kenneth Sprunt)

Azalea Festival Presidents:

Hugh Morton	1948	John Van B. Metts	1967	H. E. Miller, Jr.	1986
David Harriss	1949	E. E. (Bill) Huffine	1968	Doug Echols	1987
Hal Love	1950	F. P. Fensel	1969	David Ward	1988
E. L. White	1951	William H. Sutton	1970	William H. Cameron	1989
Allen Jones	1952	Frank Ballard	1971	Rodney H. Everhart	1990
Rye B. Page	1953	John C. Hall	1972	Lee Weddle	1991
Edward L. Ward	1954	Franklin Elmore	1973	Beverly A. Jurgensen	1992
Louis Latham	1955	W. Allen Cobb	1974	James B. (Jay) Stokely	1993
William S. Rehder	1956	W. R. Burns	1975	H. E. "Hank" Miller III	1994
W. G. Broadfoot, Jr.	1957	Thurman W. Sallade	1976	Helen Skelton Lewis	1995
Frederick Willetts, Jr.	1958	Terry Horton	1977	Michael W. Creed	1996
Walker Taylor, Jr.	1959	Russell Clark	1978	Uldis "Buzz" Birzenieks	1997
Robert A. Little	1960	Paul Burton	1979	W. Allen Cobb, Jr.	1998
E. S. Capps	1961	F. P. Fensel, Jr.	1980	Bill Rudisill	1999
Allan T. Strange	1962	Donald Britt	1981	Emily Longley	2000
L. Bradford Tillery	1963	Ed Ward, Jr.	1982	David Kauffman	2001
W. A. Raney	1964	James Carter	1983	Paula Lentz	2002
G. Stanley Rehder	1965	William N. Rose	1984	Wanda Copley	2003
Lloyd W. Moore	1966	Frederick Willetts III	1985		

Interesting bits

1939-1940 yearbook includes this anonymous prayer:

A Garden Prayer
Help us, O Lord, to grasp the meaning of happy grow-ing things…the mystery of opining bud and floating seed…that we may weave it into the tissue of our faith in life eternal.
Give us wisdom to cultivate our minds as diligently as we nurture tender seedlings, and patience to weed out envy and malice as we uproot troublesome weeds.
Teach us to seek root growth rather than a fleeting cul-ture, and to cultivate those traits which brighten under adversity with the perennial loveliness of hardy bor-ders.
Thank God for gardens and their messages, today and always.
Amen.

The 1950 yearbook includes this verse by the English poet, Cowper:

To meliorate and tame the stubborn soil;
To give dissimilar, yet fruitful land,
The grain or herb, or plant that each demands…
To mark the matchless working of the power
That shuts within its seed the future flower…

"Incidentally, Wilmington is the most northerly point where palms, palmetto, and other forms of tropical vegetation grow in the open," wrote Chamber of Commerce director Louis T. Moore in 1930.

My Day With God
By Churchill Bragaw

Oh Lord, you've been so good to me,
So patient and so kind,
If I'm not in church today,
I hope that you won't mind.
For I'll be back when the leaves are gone,
When the cold wind starts to blow.
And I'll sing your hymns 'neath somber walls
When the ground is covered with snow.

But, Lord; there's a song in the air today,
And my sould won't be denied,
I'll put it out, I said to myself,
And then in vain I tried.
So I'm off to the reeking cypress swamp
Where the ferns grow like a prayer,
And blue-bells cover the vine-clad stumps
Like jewels in a maiden's hair.

I'll hear the log-cock's drumming song
From the top of an old dead tree.
And my heart will be filled with good will to man
Through all eternity.
I'll hear the bob-white's love song go
To his mate in the grass nearby,
We'll face no pulpit there today –
Just Nature, and You and I.

The fragrance of the jasmine vine
Shall be my hourly creed.
My sermon will come from the blue above,
Where there's none of strife and greed.
My choir will be the liquid notes
Of the birds in the trees above,
And the last refrain of the final amen
Will come from the mourning dove.

So God; on Sundays when I'm not in church
And there seems so little of prayer,
You'll know, dear Lord, that I'm in the woods
And I'll always find you there.

(Churchill Bragaw was a horticulturist and manager of Orton Plantation. He was killed during World War II. His poem would have been the bane of a minister but speaks beautifully of his tender spirit.)

Acknowlegements

I am especially grateful to two members of Cape Fear Garden Club. Elaine Henson and Teresa Hill originated the idea for this book and asked me to write it. In addition, Elaine has been an energetic assistant.

The late Tabitha Hutaff McEachern, a longtime friend to the garden club, gave a generous gift towards the publishing of *Belles and Blooms: Cape Fear Garden Club and the North Carolina Azalea Festival.*

I also owe gratitude to Hugh Morton, Henry Rehder, Thomas S. Kenan III, Josephine Horton, Eli Naeher, Merle Chamberlain, Dr. James Rush Beeler, Beverly Tetterton, Katherine Meier Cameron, Joseph Sheppard, Carroll Robbins Jones, Henry B. Rehder, Ann Brennan, Betsy and Kenneth Sprunt, Eve Carr, Edwina Batson, Kenneth Davis, Tim Bottoms, Barbara Rowe, Hugh MacRae II, Nina Cain, JoAnne Mathis, Dorothy Gaither, Mr. and Mrs. George Taylor, Sue Boney Ives, Elizabeth McCauley, Ann Penton Longley, Diane Dayton, and Miriam Burns Whitford, Millie Maready, Gayle Ward, Freda Wilkins, Mary Lee Stewart, Jenene Smith, Dianne Lynch, and Mary Lou McEachern.

Jane L. Baldridge deserves many kudos. This is our fourth book together and our work is like a sort of dance. There are few things as satisfying as laboring hard with someone who is a perfect professional match.

My parents, Mr. and Mrs. Joseph Wright Taylor, Jr., were helpful. My daughters, Taylor and Catherine, are unending sources of wonder and renewal. And, above all, I wish to thank my husband, Frederick L. Block, for his encouragement, praise, and love.

Select Bibliography:

Azalea Festival Planning Schedules: 1948, 1949 (Betsy Long Sprunt Collection)

Caroline Meares Collection. Perkins Library, Duke University.

A Bird's Eye View of the Tourist Training Program, sponsored by Wilmington College and Wilmington Merchants' Association. Wilmington, N. C. No date (circa 1952)

Cape Fear Garden Club. Yearbooks from 1928 until 2003. Special Collections: New Hanover County Public Library.

Cape Fear Garden Council Scrapbooks Special Collections: New Hanover County Public Library.

Campbell, Walter E. *Across Fortune's Tracks: A Biography of William Rand Kenan, Jr.* Chapel Hill and London, 1996.

Cashman, Diane Cobb. *Cape Fear Country Club: 1896-1996.* Wilmington, 1996.

Daughters of the American Revolution notebooks: Lower Cape Fear Historical Society.

Edwards, Elizabeth. *A History of Cape Fear Garden Club.*

Fonvielle, Chris E., Jr. "Hijacked! Chris E. 'Gene' Fonvielle and James B. 'Bunny' Hines Go To Cuba." 1969.

Marie Rehder Gerdes Collection. Special Collections, New Hanover County Public Library.

Grizzle, Ralph. "Guitar Man: Arthur Smith"

Hewlett, Crockette W., and Mona Smalley. *Between the Creeks, Revised.* Wilmington, 1985.

Kenan Collection. Special Collections, University of North Carolina at Wilmington.

Parker, Constance N. *Cape Fear Garden Club: Historical Highlights of Sixty Years.* Wilmington, 1985.

Pendleton, May. *Brief Resume of the Activities of the Cape Fear Garden Club, 1951-1973,* (1955-56 yearbook). New Hanover County Public Library.

Wallis, Martha Hyer. *Finding My Way: A Hollywood Memoir.* New York, 1990.

Interviews:

Betsy Long Sprunt, Kenneth Murchison
Sprunt, Henry B. Rehder, Hugh MacRae, Hugh
Morton, Josephine Corbett Horton, Frederick
L. Block, Elsie Corbett Hatch, Catherine Meier
Cameron, Walker Taylor III, Joanne Corbett,
Allan T. Strange, Jo Chadwick, Elaine
Henson,Thurston Watkins, Jr., Marth Blacher,
Dr. Harry Van Velsor, Gayle Ward, Hannah S.
Block, Millie Maready, James Knowlton
Wright, Penelope Spicer-Sidbury.
Author's correspondence with:
Sara Shane (Elaine Hollingsworth), Joan Van
Ark, Martha Hyer Wallis,
Materials from New Hanover County Public
Library, Lower Cape Fear Historical Society,
Cape Fear Museum, UNCW, Edwina Batson
Azalea Festival Collection. Betsy Long Sprunt
Collection.

Chicago Tribune:
April 10, 1966

News and Observer:
March 27, 1955
April 9, 1962
April 6, 1970

Wilmington Star News:
December 12, 1948
March 28, 1948
Dec. 12, 1948
March 27, 1949
April 1, 1949
March 22, 1953
March 10, 1954
March 26, 1955
April 2, 1955

April 3, 1955
April 5, 1959
June 3, 1962
March 15, 1964
May 17, 1964
April 4, 1965
March 22, 1966
April 22, 1966
March 31, 1967
April 14, 1969
April 5, 1970
December 19, 1971
April 17, 1973
March 24, 1974
April 13,1975
April 6, 1979
April 10, 1981
March 5, 1986
April 11, 1987
April 7, 1988
April 13-14, 1989
April 3, 1990
April 11-12, 1991
April 10, 1992
April 11, 1992
April 1, 1993
April 11, 1996
January 31, 1998
April 4, 1999
April 7, 2000
August 6, 2003

Videos:
1948 Azalea Festival. Written by Hugh
Morton, narrated by Ted Malone. Hollysmith
Productions of Charlotte. (Cape Fear
Museum)
1953 Azalea Festival   (Cape Fear Museum)
April 11, 1992: Hugh Morton slide show of the
Azalea Festival (Cape Fear Museum)

Select Index of Gardens and Individuals:

Leaving Union Station: Mr. & Mrs. Ted Malone and Jacqueline White in 1948. (Photo by Hugh Morton)

Dedication page identifications and credits: (Clockwise, from top left, with photographers noted in parentheses) Azalea Belle (Millie Maready), Venus flytrap (Hugh Morton), Azalea Tour volunteers Christiane Dybvik and Barbara J. Sillars (Millie Maready), the McMurry Garden, 2001 (Millie Maready), Azalea Belle (Millie Maready), Azalea Belles (Millie Maready), Azalea Belle Jennifer Ward and her mother, Belle sponsor Gayle Ward, 1991 (David Ward), Azalea Belle (Millie Maready), Azalea Belle with frog sculpture at the Mildenberg Garden, 2002 (Millie Maready), six belles a-kicking at the McEachern house on Cedar Island (Gayle Ward), Azalea Belles at the Queen's Welcome, Greenfield Lake, 2003 (Millie Maready), Azalea Belle (Millie Maready), Amanda Gresham in the Overton Garden, 2003 (Millie Maready), a pink perfection (Gilliam K. Horton, courtesy of Josephine Corbett Horton), Azalea Belle Sarada Patterson (Edwina Batson).

Kelly Ripa and Henry Rehder. (Henry B. Rehder Collection)

WILMINGTON
NORTH CAROLINA

AZALEA
FESTIVAL

WILMI
NORTH C

AZA
FESTI